D1379748

MARIA MITCHELL AND

THE SEXING OF SCIENCE

Maria Mitchell and the Sexing of Science

An Astronomer among the American Romantics

RENÉE BERGLAND

Beacon Press
BOSTON

Beacon Press
25 Beacon Street
Boston, Massachusetts 02108–2892
www.beacon.org

Beacon Press books
are published under the auspices of
the Unitarian Universalist Association of Congregations.

11 10 09 08   8 7 6 5 4 3 2 1

This book is printed on acid-free paper that meets the uncoated paper
ANSI/NISO specifications for permanence as revised in 1992.

Composition by Wilsted & Taylor Publishing Services

Library of Congress Cataloging-in-Publication Data

Bergland, Renée L.
 Maria Mitchell and the sexing of science : an astronomer among the
American romantics / Renée Bergland.
   p. cm.
 Includes bibliographical references and index.
 ISBN 978-0-8070-2142-2 (acid-free paper)
 1. Mitchell, Maria, 1818–1889. 2. Women astronomers—United States—
Biography. 3. Astronomers—United States—Biography. I. Title.

QB36.M7B47 2008
520.92—dc22
 [B]   2007035133

Frontispiece of Maria Mitchell from a photo by William Summerhayes,
circa 1855. Courtesy of the Nantucket Maria Mitchell Association.

DEDICATED TO ANNELISE

CONTENTS

INTRODUCTION: Venus in the Sunshine *ix*

CHAPTER 1   Urania's Island   *1*

CHAPTER 2   Nantucket Athena   *21*

CHAPTER 3   The Sexes of Science   *42*

CHAPTER 4   Miss Mitchell's Comet   *53*

CHAPTER 5   "A Center of Rude Eyes and Tongues"   *72*

CHAPTER 6   The Shoulders of Giants   *91*

CHAPTER 7   The Yankee Corinnes   *115*

CHAPTER 8   A Mentor in Florence   *137*

CHAPTER 9   The War Years   *153*

CHAPTER 10   Vassar Female College   *166*

CHAPTER 11   No Miserable Bluestocking   *177*

CHAPTER 12   "Good Woman That She Is" *192*

CHAPTER 13   The Undevout Astronomer *211*

CHAPTER 14   Retrograde Motion *224*

CHAPTER 15   Urania's Inversion *239*

EPILOGUE *250*

ACKNOWLEDGMENTS *260*

NOTES *265*

INDEX *284*

By 1879, Maria Mitchell was such a celebrity that people who sat next to her at a meal or glimpsed her across a train platform often wrote to their hometown newspapers to report the sightings. That year, Lilla Barnard saw the famous astronomer at a large public dinner in New York City. Barnard was from Mitchell's home, the island of Nantucket, and she remembered that Mitchell had been very kind to her when she was a girl. She presented the sixty-year-old with a rose and then hurried home to write an article for *The Woman's Journal* about her brush with the national heroine. In the article, Lilla Barnard remembered that back in 1861, "our astronomer, even then known to the whole world," had welcomed her into her private observatory with a flashing smile.

In the summer of 1861, Nantucket was empty. A few disused whaling ships clanked against the piers in the harbor, but no sailors roamed the streets, and many of the houses and shops were shuttered. Lilla Barnard trudged up Nantucket's sandy Main Street, looked restlessly into shop windows, and listened to the gulls shrieking and the flags flapping in the endless breeze; the hours after the library closed felt infinitely long. Lilla spent mornings at the Nantucket Atheneum, a grand and welcoming Parthenon of a place with heavy, creaking wooden doors between its soaring white wooden columns. When Lilla was learning to read, the Nantucket Atheneum had seemed like a temple, and the librarian at that time, Miss Mitchell, had seemed almost like a goddess. As a very young girl, Barnard had thought Miss Mitchell was Nantucket's own Athena; after all, she ran the Nantucket Atheneum, dispensing books and wisdom to everyone who entered.

But the summer of 1861 was different. Miss Mitchell was no longer the librarian; instead, she came to the Atheneum in the mornings to read, study, and feed her restless curiosity, just as Lilla did. Sometimes Lilla sat at the same long table and tried to assume some of the scholarly dignity of the quiet Quaker woman in the simple black dress. This afternoon, Lilla perked up a little when she saw Miss Mitchell walking briskly up the street. Unlike most people on the island, Miss Mitchell was interesting. One of Mitchell's distant cousins, Benjamin Franklin, had discovered electricity, and Lilla secretly thought Miss Mitchell looked a bit like a skinny Ben Franklin in a dress. She had overheard her mother's friends say that there was something electric about Miss Mitchell, and Lilla agreed. Maria Mitchell crackled and sparkled somehow, even when the rest of the island was torpid and sleepy in the sunshine.

Miss Mitchell flashed the young girl an encouraging smile. "Good afternoon, Lilla," she said. "Come with me and I will show you Venus while the sun shines."

"Her quick spoken words fell on glad ears," Barnard recalled decades later, "and the child sprang after her, believing, because Miss Mitchell said so, but in utter bewilderment as she looked into the sunny sky."[1] Mitchell took Lilla's hand and half dragged her up the street, not slowing down to match the child's pace. Lilla loved the jolt of energy she got from holding Miss Mitchell's hand and scurried to keep up, unaware that she was mimicking her idol's energetic stride.

Miss Mitchell had a small observatory located behind the Coffin School on Winter Street. Her telescope was one of the finest in the country, and, gloriously, it had been given to her by all the "women of America." *Emerson's United States Magazine* had run a campaign urging girls and women to send in whatever money they could spare to help buy the great astronomer a great telescope. When the telescope had been installed, an article in *Scientific American* had documented the construction of the observatory. Now, Lilla was finally being invited into the little round house.

The telescope was much longer than she had expected: one end almost touched the dome, twelve feet above. Miss Mitchell checked some papers and then gave the rotating dome a shove to spin its opening toward the sun. The cannonballs on which the roof rested rolled smoothly. She quickly adjusted the instrument, pulled a stool toward the eyepiece, and invited Lilla to take the first look. Years later, Barnard recalled that in her first encounter with a telescope, she was awed by the cold brilliance of the crescent-shaped planet in the bright blue sky and encouraged by the warm brilliance of Miss Mitchell's countenance.[2] Barnard remembered the "warmth and depth of Maria Mitchell's affectionate nature" and her "whole-souled generosity" at least as well as she remembered the glimpse of Venus shining in the daylight.[3]

Venus is the brightest of the planets—in our sky, only the sun and the moon exceed its brilliance. It is not hard to spot the planet during the day, if you know where to look. And Maria Mitchell knew where to look: one of very few professional astronomers in the United States before the Civil War, Mitchell was employed by the United States *Nautical Almanac* to calculate the orbit of Venus. From 1849 to 1868, she was the "computer of Venus." The *Nautical Almanac* was based on mathematics rather than observation; Mitchell did not need to observe Venus in order to predict its motion. Instead, she computed pages and pages of calculations, eventually reducing them to charts that accurately predicted the position of Venus in the sky years in advance. American sailors all over the world used the planetary charts in the almanac for celestial navigation, while astronomers used them to calibrate their own observations; Venus was a particularly important reference point because it was large and easy to see. The U.S. government employed just eleven elite astronomer-mathematicians in this capacity. Benjamin Pierce, a Harvard professor of mathematics and the inventor of binary arithmetic, was the project's chief mathematician. Sears Cook Walker computed the position of Neptune for the almanac (he was the first person to succeed in calculating Neptune's orbit, an accomplishment that was one of the greatest feats of nineteenth-century astronomy). John Runkle, who would one day be president of the Massachusetts Institute of Technology, helped calculate the motions of the moon.[4] Mitchell was the only woman in the group. Charles Henry Davis, director of the *Nautical Almanac,* had matched her with Venus as a bit of joking gallantry: "As it is 'Venus who brings everything that's fair,' I therefore assign you the ephemeris of Venus, you being my only *fair* assistant."[5]

In 1861, Mitchell was one of the most famous women in the world, but she was modest by nature and disliked publicity. Upstairs in her bedroom above the Pacific Bank, Mitchell had a medal that had been awarded to her by the king of Denmark for her discovery of the comet that had brought her international renown. The medal was a three-inch disk of solid gold with Miss Mitchell's name engraved on it; its dull gleam bespoke the solid weight of her accomplishment. On the medal's reverse was a finely detailed relief of Urania, the muse of astronomy, upon her orb.

It is highly unlikely that Mitchell showed her medal to Lilla Barnard; that would have been showing off, and Maria Mitchell couldn't abide a braggart. But if she had shown it to anyone on Nantucket, it would have been to a talented girl. Maria Mitchell was an enormous success as an astronomer, but educating girls in the sciences was even closer to her heart than astronomy. As she grew older, Mitchell learned to put aside her own personal shyness so that she could serve as a role model for young women interested in learning.

Lilla found it inspiring to work beside Mitchell in the small observatory: to see how deeply she could concentrate; to be given small jobs that were actually useful; and to find herself doing real science. When Miss Mitchell told Lilla that the research with which she was assisting would be published in the Nantucket *Inquirer and Mirror,* it must have seemed to Lilla that her childhood impression was right: Miss Mitchell really was the gray-eyed goddess Athena leading a young acolyte to wisdom. Miss Mitchell certainly didn't fit Lilla's ideas of Venus; she was far too angular. But she was the *computer* of Venus—an astronomer, which linked her to Urania, the muse of astronomy.

In the 1860s, another goddess rose to challenge Athena, Venus, and Urania. Torn by the violence of Civil War, Americans were obsessed with the idea of quiet domesticity to counteract the tumult outside. In 1861, the prevailing womanly role model was not a goddess but an angel, the self-effacing Angel in the House who gently cleaned and cooked and organized the household into an oasis of peace and quiet. Lilla might have known that Maria Mitchell was the only daughter left at home, and that her mother was very sick. She might have been aware that most of Mitchell's time was spent quietly sitting with her mother, soothing her, trying to play the role of household angel that circumstances had thrust upon her. But Mitchell was no Angel in the House—domesticity was an awkward fit for the Nantucket Athena, the computer of Venus, the American Urania.

When we look up at the night sky, the stars seem to move in a majestic progression, rising from the eastern horizon and slowly traversing westward, circling the pole star. In reality, the stars stand still; the earth spins, and thus our view of the stars changes. Only a few shining points in the sky change position among the others: the wandering planets, whose orbits around the nearby sun are close enough to us that we can watch their travels.

Mercury and Venus appear to circle the sun because they are closer to it than the earth is. We can see them at dawn and dusk, but their positions, unlike those of the stars, are hard to predict. Sometimes they seem to move backward, from west to east rather than east to west.

For many centuries, navigators relied on the positions of the stars and the planets to determine their own positions on the earth, and they relied on predictions of future positions to plan their routes.

Calculating planetary motion is a delicate problem, as one must take into account the sun's motion; the earth's daily rotation on its own axis; the earth's year-long elliptical revolution around the sun; and the elliptical orbits of the planets themselves. As we watch it from earth, a planet on the more pointed end of its oval-shaped orbit seems to move quickly, while on the flatter side of its orbit it seems to slow down. In Ptolemy's day, astronomers produced elaborate, flower-shaped, Spirograph-style drawings that showed Venus's forward and backward motion through the night sky. Mitchell, a more modern astronomer, offered tables instead, but these were not as different from the Ptolemaic drawings as one might imagine. Venus still wandered.

Traditionally, Venus is viewed as the female planet. Her erratic path through the skies is an apt metaphor for women's circuitous path toward political and educational rights, and at times, the story of women's rights in the United States has also moved backward. In the glorious years of Mitchell's youth on Nantucket and up to the middle of the nineteenth century, women's access to education and to the political process expanded rapidly. But this quick expansion was followed by an equally swift contraction. Even women's education—previously regarded as a clear social good—was controversial by century's end.

Women's relationship to the sciences changed most drastically of all. When Mitchell was young, science was considered a ladylike avocation, and girls were actively encouraged to study it. Mitchell never intended to challenge scientific or social conventions by her work in astronomy; on the contrary, the community supported her interests. But later generations of female students would not meet with the same encouragement. Although Mitchell is often named as one of the first American woman scientists, later in her life she feared that she might be the last. After her election to the American Academy of Arts and Sciences, in 1848, no woman followed her into that august institution during her lifetime.[6] In 1881, in a report to the Association for the Advancement of Women, Mitchell wondered, "At what time did scientific associations close to women?"[7] Her plaintive question showed that the story of her female scientific contemporaries had become a tale of retrogression—backward motion, like that of Venus—and she mourned the closing of a window of opportunity that had opened all too briefly in the early decades of the century.

Today, it is commonplace to associate science and math with men and link art and culture with women. But this segregation is more rigid for women; male artists do not surprise us as much as female scientists do, and for many, the concept of the woman scientist is stranger and more paradoxical than the idea of the male artist. Historian of science Sandra Harding has commented that "women have been more systematically excluded from doing serious science than from performing any other social activity."[8] What makes this exclusion unusual is that it happened very suddenly, in a specific period. In the first decades of the nineteenth century, science was identified more closely with girls than with boys, because it was seen as apolitical. In the 1860s, science was gradually redefined: no longer safe, it became highly speculative, even radical. Michael Faraday, Charles

Darwin, and James Clerk Maxwell offered experiments, theories, and equations that didn't necessarily accord with the accepted Protestant Christian cosmogony. Some science challenged Christian doctrine, and other work simply ignored it, but in any case science was no longer the place to learn docile obedience and respect for hierarchy. By 1873, Edward Clarke of Harvard Medical School was arguing that science itself could prove that women were physiologically unable to study it, and that those who studied too hard risked becoming "thoroughly masculine in nature, or hermaphroditic in mind."[9] Only eight years earlier, in 1865, Maria Mitchell had been appointed the first professor of astronomy at the newly opened Vassar College for women, and there had been no public murmurings about the masculinity or the hermaphroditic nature of women scientists. But by 1875, that idea had started to prevail. This was a sudden and dramatic shift. There is no parallel story in the arts of an institutional shift in gender association, nor of men suddenly being barred from the humanities. With rare exceptions, society is not troubled or threatened by the figure of the male artist.

Early in the nineteenth century, the word "scientist" didn't have masculine associations; in fact, it was coined to describe a woman. In 1834, the Cambridge don William Whewell wrote a complimentary article about Mary Somerville, a Scottish researcher whose erudite books brought together previously disparate fields of mathematics, astronomy, geology, chemistry, and physics so clearly that the texts became the backbone of Cambridge University's first science curriculum. He called Somerville a scientist, in part because "man of science" seemed inappropriate for a woman, but more significantly because Somerville's work was interdisciplinary. She was no mere astronomer, physicist, or chemist, but a visionary thinker who articulated the connections among the various branches of inquiry.[10] According to Somerville's biographer Kathryn Neeley, Whewell's coinage of the word "scientist" was not meant to be merely a gender-neutral term. Whewell wanted a word that actively celebrated "the peculiar illumination of the female mind": the ability to synthesize separate fields into a single discipline.[11]

In the early nineteenth century, women scientists were seen as less threatening than other women professionals. In 1848, Lucretia Mott argued for women's political rights in the "Discourse on

Woman" that she delivered at Seneca Falls and around the country. She started gently, mentioning women's achievements that she knew would not upset her listeners. "Do we shrink," she asked, "from reading the announcement that Mrs. Somerville is made an honorary member of a scientific association? That Miss Herschel has made some discoveries, and is prepared to take her equal part in science? Or that Miss Mitchell of Nantucket has lately discovered a planet, long looked for?"[12] Mott knew that her audience would not shrink from the idea of women in science but rather would embrace it. A woman scientist was the most socially acceptable female role model she could think of, and Maria Mitchell was the American woman scientist whom she trusted every listener would like.

John Raymond, the first president of Vassar College, was well known for his stance against women's political rights, but he shared society's assumption that science was suitable for females, and he was a firm friend to Maria Mitchell. Addressing a national convention of Baptists, Raymond said,

> In many of the processes of the laboratory, in the arrangement and care of great collections, in the keeping of minute and voluminous records, in difficult and delicate computations, and in like work of which there is so much to be done in chemistry, astronomy, and in the whole range of natural history, and on the manner of doing which so much is often depending, one thoroughly trained woman is often worth any number of young men, who, with rare and womanly exceptions, cannot do such work well if they try, and would not want to if they could.[13]

For President Raymond, studying science was "womanly," safely outside the potentially dangerous ideological realms of law or history or theology. Those realms, he was sure, would not suit women:

> I have said nothing of the bar, the bench, or the pulpit, and nothing of the arena of public political discussion and conflict, into which admittance for women is so clamorously demanded by some, because I am more than doubtful whether, as these are at present organized, woman has any vocation to either, and because I believe, for a long time to come, highly and truly educated women may find more congenial and less questionable fields for the employment of their powers.[14]

All of which is to say that in 1861, when Maria Mitchell and Lilla Barnard were in the observatory together, this seemed a natural place for a woman and a girl to be. Nantucket was at the forefront of women's education—Nantucketers had long believed that girls should have the same educational opportunities as boys—and if girls were to study, there was no more appropriate subject than astronomy.

With the benefit of hindsight, we can see the 1860s as a turning point in the history of the scientific education of girls and women. It is particularly appropriate that Maria Mitchell showed Lilla Barnard the wandering planet Venus, because women would very soon be pulled backward by immensely powerful, almost gravitational, social forces. A few years after that, one of Mitchell's students at Vassar wrote a silly poem about what she had learned in Mitchell's astronomy class:

> Of all the great iconoclasts
> The mightiest one is knowledge
> And many fancies are destroyed
> By every course in college
>
> No more we look with simple gaze
> upon fair Venus bright
> Nor Jupiter the mighty one
> Nor Mars with ruddy light
> Without recalling to our minds
> Their motion retrograde.

Mitchell's students' illusions were shattered by what she taught them, especially as the 1860s ticked toward the even more sexist 1870s and 1880s. One of the most enduring lessons they learned was that Venus, the female planet, was often pulled into retrograde motion.

In the nineteenth century, as now, most people liked the idea of progress. Retrograde motion was a disturbing concept. It was much more comfortable to believe that conditions for women were steadily improving and would continue to do so. But one problem inherent in a historical narrative of progress is that we can overlook lost opportunities. Despite Maria Mitchell's enormous accomplishments and justified renown, her story is sad: in the last thirty years of her life, the support of women's education and the prospects for women in science steadily regressed. Mitchell was angrily aware that the situation was worse for her Vassar students than it had been for her, and that her students' daughters and nieces would face even greater educational and professional barriers.

Still, there is great hope in Mitchell's story. It is a happy fact that Maria Mitchell's career was possible, and it is heartening to know that there was a time in America, however brief, when girls were strongly encouraged to pursue the sciences. We currently live in an era when the bias against women in science seems an eternal constant. It is exciting to think about a time in our own culture when things were quite different, when Lilla Barnard and Maria Mitchell strode purposefully up Nantucket's Main Street, hand in hand, on their way to Mitchell's observatory.

CHAPTER I

Urania's Island

Before Maria Mitchell was born, there were two lending libraries on Nantucket and many collections of books in private homes. Lydia Coleman, later Maria Mitchell's mother, was said to be the only person who had read every book on the island. William Mitchell remembered of his wife: "Her form was perfect in its proportions, rather tall and slender, and early, as in later life, she was very upright. Her step was always short, and her motions quick. Her face was not what would be called handsome. Her features were well-formed, but her skin was slightly freckled. Her eyes were her commanding feature. It was in these that the great qualities of her mind and heart could be read.... White dresses were evidently her prevailing taste while young, and in these she appeared as elegant in person as beautiful in form. She was an intense reader."[1]

Curious, sharp-witted, and largely self-taught, Lydia Coleman Mitchell went to work in both libraries so she could read their stock of books. She was a close relation of Walter Folger and his sister Phebe, and therefore had the run of the extensive library in Walter's home. Before her marriage, Lydia worked as a teacher in a Quaker school, an appointment that signified the Quaker meeting's high opinion of her knowledge. A passionate student with wide-ranging interests, she particularly loved reading fiction. When she was a young woman, her favorite author was Maria Edgeworth, who wrote novels, educational treatises, and educational stories that offered moral guidance along with strong doses of practical mathematics and the sciences. Maria Mitchell was named at least in part because of her mother's love for Maria Edgeworth, whose works were a constant resource in Lydia's first years of parenting.

William Mitchell was Lydia Coleman's intellectual equal. He was one of the few island boys who had won a place at Harvard, and he planned to enroll when he turned nineteen. But when the time came, he later recalled, "I found I could not leave Lydia Coleman."[2] If Lydia could have gone to Harvard, or indeed to any college, they both would surely have gone on to higher education, but in 1811, college was not an option for Lydia. And so they married that year and moved out to a small cottage near 'Sconset Bluffs. A year later, they moved to Vestal Street in Nantucket town. Their address was a happy coincidence: it was a family property. But since both Lydia and William were strong supporters of education for girls and women, it was fitting that they lived on a street named for the vestal virgins. In 1812, "vestal" had two different connotations. Classically, Vesta was the ancient Roman goddess of the hearth, and the vestal virgins were chaste women who dedicated themselves to the study of higher things. Scientifically, Vesta was the name of a very bright asteroid discovered by a German astronomer in 1807. Both meanings were significant for the Mitchells, who were at home with classical goddesses *and* nineteenth-century astronomy.

By all accounts, the Mitchells' marriage was remarkably happy. The first winter was a hungry one, partly because the War of 1812 blockaded the ports, and the next fifteen years were lean as well, largely because the family grew so quickly: Lydia and William had ten children, nine of whom survived to adulthood. Maria was born on August 1, 1818, the Mitchells' third child and second daughter. The oldest child, Andrew, was born in 1814; next came Sally in 1816. After Maria, more children followed at regular intervals: Anne (1820), Francis Macy (1823), William Forster (1825), Phebe (1828), the twins, Henry and Eliza (1830). Henry's twin, Eliza, died very young; the last Mitchell sibling, Eliza Catherine, who was known as Kate, was born in 1833.

In a family of quick wits, Maria's were the sharpest. Her sister Phebe recalled that:

In all amusements which required quickness and ready wit, she was very happy. She was very fond of children and knew how to amuse them and take care of them. As she had half a dozen younger brothers and sisters, she had ample opportunity to make herself useful. She was a capital storyteller, and always had a story on hand to divert a wayward child, or to soothe the little sister who was lying awake, and afraid of the dark. She wrote a great many stories, printed them with a pen, and bound them in pretty covers. Most of them were destroyed long ago.[3]

Maria was always willing to pitch in with the household chores, but much of the work of the Mitchell household was brain work. "We always had books, and were a bookish people," she later recalled.[4] The family's house was so tiny and their possessions so few that physical drudgery was limited to the seemingly endless rounds of washing up. Beyond that, Maria read novels aloud to her mother and amused her younger siblings with stories, helped them with letters, and led them through the pages of Colburn's *Algebra*. The Mitchell family algebra book is still preserved in the archives of the Maria Mitchell Association, with the names of eight Mitchell siblings, starting with Sally, listed in succession on its flyleaf. Perhaps the oldest child, Andrew, had run away to sea (as a ship's boy on his uncle's whaler) before the family funds stretched to purchase an algebra book.

Studying math was vitally important to the Mitchells because the central work of the family was astronomy. Unless the clouds prevented him, William Mitchell spent every evening observing the skies. His daughter Phebe recalled that "to the children, accustomed to seeing such observations going on, the important study in the world seemed to be astronomy. One by one, as they became old enough, they were drafted into the service of counting seconds by the chronometer during the observations. Some of them took an interest in the matter at hand while others considered it rather stupid work, but they all drank in so much of this atmosphere that if anyone had asked a little child in this family, 'Who was the greatest man that ever lived?' the answer would have come promptly, 'Herschel.'"[5]

The British astronomer William Herschel (1738–1822) was a family idol, as was his sister Caroline, who worked as his assistant and

discovered eight comets on her own. Phebe's recollections make it
clear that the Mitchells hoped to emulate the Herschels. They were
an astronomical family.

Lydia Mitchell put much of her energy and practical skills into running the household; difficult, since they were a big family with a
small income. She taught her children herself at the long kitchen
table: letters, numbers, and challenging word games interspersed
with silly poems, puzzles, and stories. With so many children, the
house was like a crowded little school. In the years before standardized education, few schools stayed open longer than a year or two.
Capable women, and sometimes men, often opened their homes to
a few paying students who came for a few hours each morning. The
Friends' meeting often sponsored a school for a term or two, but
most were short-lived. The town debated strategies for sponsoring its
own schools, but until the Coffin School opened its doors in the late
1850s, none of the town schools lasted much longer than the private
ones. The Mitchells took advantage of whatever schooling opportunities they could find for their children, and when Nantucket finally
funded a large common school, William served as its first schoolmaster. Unfortunately, William's involvement with the town school
lasted only a few years. Lydia and William supervised their children's
learning much more closely than they might have if Nantucket
had had an established school. William spent far more time at home
than most Nantucket men did; he never went to sea, and when the
children were young, he frequently worked at home or in situations
where he could bring the children along with him.

The Mitchell children were not only encouraged to learn, they
were also surrounded by an atmosphere of constant inquiry and intellectual exploration. Suspended from the Mitchells' kitchen ceiling, a large crystalline bowl of water caught the slanting sunlight
and threw rainbows around the room. The whole family probably
enjoyed the bright play of color on the plastered walls, but their
somber, achromatic Quaker culture would have discouraged them
from talking much about that pleasure. What they could talk about
was optics: angles of light and refraction. And so, in the guise of scientific experimentation, the children watched color play brightly
through their household. As would-be physicists, they experimented

with prisms and crystals and rainbows. As aspiring botanists, they grew banks of wildly colorful flowers all around the cottage. As astronomer's helpers, they sometimes stayed up late at night so they could climb out on the roof and gaze at the sky.

Thanks to Lydia Mitchell, the children knew by heart the stories of Maria Edgeworth, from which they learned about narrative and about practical applications of knowledge. But William Mitchell's passions were visual: he loved color, and he loved to look at beautiful things. Astronomy combined his philosophical interests with his love of gazing into infinite space. Although he never made much money at it, William consistently found astronomical work to do along with his other jobs. Many years later, Anne Mitchell recalled the morning when her sister Maria was singled out to become William's assistant, just like the world-famous Caroline Herschel had been.

> One morning we were preparing for school, and were about ready to start, when my father put his head into the room and said: "which one of my children will count seconds for me? The quicksilver is ready." The quicksilver found an artificial horizon, as I learned 20 years after. Maria knew it then. No one replied to my father's question, but my mother, who had long since discovered the inclination of my sister's remarkable powers, and knew very well what direction her future studies were to take, said quietly as she looked at Maria: "Thee is the one to help father."
>
> Maria readily drew off her mittens and went to him. My father at that time had the chronometers of the 95 ships which composed our large whaling fleet, in his hands, as they were from time to time brought into port. His observations supplied the rate for the next voyage. Maria began this morning to help my father, by counting seconds, and from that day continued as his assistant, finally rating them herself as accurately as her teacher. She was but 11 years old, and from that time, her studies were never interrupted.[6]

Maria Mitchell's birthplace on Vestal Street in Nantucket is an impossibly sweet little house. There is something charmingly doll-like about its perfectly symmetrical gray-shingled facade, its white picket fence, and its riotous flowerbeds. On each of the two main floors, generous fireplaces opened onto two good-size rooms on opposite corners of the house. The front parlor was solemn, often cold in spite of its south-facing windows. A grandfather clock ticked industriously along, offering some comfort and company, but the parlor was never as busy or as warm as the kitchen at the back, where the Mitchells cooked and read and studied around the large open fire. Most of the time, there was no wood to spare for any but the kitchen fire, but the house was so small and the chimney so central that one fire was sufficient. The mood of the house was lively; although the Mitchells did not have much money, there was no feeling of poverty or deprivation. Despite the cramped quarters, there was always room for a new book or an interesting seashell, rock, or feather.

Upstairs, the floor plan was basically the same as the downstairs. There were two larger bedrooms with fireplaces and bedsteads, a smaller room at the back corner, and a hallway and stairs to the attic. In such a small house, it was frustrating that the stairs took up so much space. One winter, William decided to reframe the stairs to create a little more space; downstairs he added a pantry off the kitchen toward the front of the house. Upstairs, two of the pretty front windows wasted the warmth and brightness of their southern exposure on the hall by the attic stairs, so William carefully framed a tiny room around the corner window at the foot of the stairs, with a built-in desk and a few shelves on the wall above. The woodwork was simple and elegant, gleaming with fresh white paint. There was a little over a square foot of floor space—barely enough room to shut herself in with a straight-back chair—but Maria Mitchell finally had a quiet place to work on her math. It was the perfect place to keep her sextant and her small collection of mathematical texts. She made a little do-not-disturb sign—MISS MITCHELL IS BUSY. DO NOT KNOCK—and hung it on the door whenever she needed to concentrate away from the rough-and-tumble of her many siblings and their friends. Miss Mitchell was only thirteen, but her parents had found a way to give her a room of her own.

William made the effort to carve out a study space for his daughter not only because she was desperate for a chance to study, but also because she could be a bigger help to him if she was trained. Her assistance during his observations was invaluable. He regretted his own relative ignorance where higher math was concerned, and he knew he would never have the time to learn advanced mathematics; depending on the season and the year, he had barrels to build, candles to mold, maps to make, reefs to chart, meetings to attend, or schools to run.

The narrow roof walk atop the Mitchells' house was just a few boards wide. The house was small and old, and the roof pitched steeply down and away from the little platform. Maria loved to climb from the garret up through the trapdoor and onto the little walkway by the chimney. The roof seemed incredibly high for such a small house. Up there, with just one railing at the level of her knees, young Maria felt she had climbed into the sky itself. When it was windy, the small trees of the neighborhood tossed below her and the waves slapped and crashed in the distance, punctuated by buoy bells and raucous seagulls. When the weather was calm, Maria felt as if she was floating high above the island and the sea, in the infinities of space.

Maria Mitchell loved the night sky. She felt real affection for the stars and planets, as much for their infinitely variable beauty as for the elegant geometry and fascinating mathematics of their grand and silent music. Once, after a few weeks of cloudy skies, she wrote,

> I saw the stars in the evening of the tenth and met them like old friends from whom I had long been parted. They had been absent from my eyes three weeks. I swept round for comets about an hour, and then I amused myself with noticing the varieties of color.... The tints of the different stars are so delicate in their variety and the grouping has all the infinity of a Kaleidoscope, infinitely extended.[7]

Maria also loved the solitude of the roof. After all, the little house was crowded: with nine children in all, three or four would sleep head to toe, sardine-style, in a small bed. Although Maria found the peace that she craved on the roof, she also found companionship, both with the stars and with her father. William Mitchell was a kind and attentive man, but as a freelance astronomer, a jack-of-all-trades, a sometime schoolmaster (with 202 students in his busiest year), and the father of nine, he could be hard to pin down. On the roof, Maria had his undivided attention, and so it was partly for her father's sake that she fell in love with astronomy.

In the autumn of 1828, when Maria Mitchell was ten, William

offered a series of public lectures on astronomy. Maria's father was a great teacher, and he was especially beloved on the island for his astronomy talks. In the *Nantucket Inquirer* of October 18, 1828, Samuel Haynes Jenks wrote a glowing review of Mitchell's lectures. Jenks commended the "plain and perspicuous style adopted by Mr. Mitchell," going on to say that

> he knows what language constitutes the most perfect eloquence—that which conveys to an audience a clear and full conception of the subject on which the speaker addresses his hearers. In this particular, Mr. Mitchell is remarkably happy; and his lucid explanations, illustrated by diagrams &c. prove that his mind has often traveled far and wide over those astral regions which form the spangled canopy of evening.

William Mitchell had a certain flair for presentation. The main reasons that his lectures were "remarkably happy" were his wonderful diagrams, models, and demonstrations. Maria's younger sister Anne recalled that her father "when a young man had given courses of lectures on his favorite topics; this was long . . . before there were orreries, or any of the modern facilities by which to illustrate the phenomena of the heavens. Consequently, his diagrams were homemade; . . . [he used] enormous balls of hard wood, . . . to illustrate axes, poles, etc., of the different worlds; these spheres were in some cases six inches in diameter, in others a foot."[8]

When Mitchell demonstrated the motion of the solar system with his wooden spheres, he needed help. Imagine him holding the sun high while directing seven children to pick up a planet apiece and walk around him in elliptical orbits. How long did it take before this stately demonstration dissolved into a wild game? Planets must have clonked against each other with the satisfying chock of hardwood; the children must have laughed. Quaker discipline did not allow dancing, but it readily made room for celestial mechanics.

For purely practical reasons, there was a large audience for William Mitchell's astronomy lectures. Nantucket was a maritime community, and celestial navigation was a vital subject for sailors and ships' officers to master. Attentive mothers needed to know fundamental astronomy in order to set their sons on the path to prosperity. They taught their daughters too, in part because Nantucketers were committed to equal educational opportunities for girls; in part because daughters would need to teach children of their own; and in part because winter nights were so long and monotonous on this flat, isolated island that astronomy was a form of entertainment. The same article that praised William Mitchell for his style also argued for the importance of astronomy as an academic subject and the particular importance of encouraging women to study it:

> As astronomy is very properly considered an arithmetical or mathematical science, too many are prone to attach to it the idea that it is quite uninteresting and that any knowledge of it, to people of the common walks of life, is altogether useless.... The plea of some, in justification of their inattention to this subject, is founded upon the absurd opinion that no *man* should enter on the study of astronomy with any other view than that of being a teacher of science or an *almanack maker;* and that it is altogether an unsuitable subject to engross the attention of women. To those who are disposed to reason in this way, (if it can be called reasoning), we will propose a few questions. If it is unnecessary for *all* to know how to calculate the eclipse of the sun or moon, is it justifiable ignorance in *any* adult person of the present enlightened age not to know the *cause* which produces that phenomenon in nature? If widening the sphere of knowledge has a direct tendency to enhance the happiness of the recipient, why not have the fountain flow as liberally for the female part of our species as for the "rougher sex"? That portion of the great family of man is certainly as important as the other, and the heavens spread as brilliant a canopy to attract the mild eye of woman as they exhibit to the sterner gaze of man. Are the imaginations of women less vivid than men? If not, why should their minds be denied the privilege of contemplating the countless orbs of argent light that roll in silent magnificence through the deep illimitable expanse?[9]

No doubt William Mitchell was pleased by his friend Jenks's article. Ten-year-old Maria was capable of reading it too, but Jenks's argument must have seemed blindingly obvious to her. In an early biography, Julia Ward Howe reported that a "sister of Miss Mitchell remembers that...Mr. Mitchell never recognized any distinction of sex in the education of his children. Maria had therefore the same education with her brothers, and was especially taught navigation. Her sister bears testimony to her persistence in study, and also to the faithfulness with which she performed her part of the work of the household, not in the shape of cake and custard making, but of solid work."[10]

Astronomy was useful work, but for the Mitchells it was always much more than that. No nineteenth-century astronomer could do much without an assistant to take notes; the contrast between the blackness of the sky and the whiteness of the page made it impossible to see the stars and take reliable notes at the same time. Working as a team, the Mitchells could see millions of miles into space, and William Mitchell had a way of making the infinities of space and time feel like a playground. In later years, Ralph Waldo Emerson, Frederick Douglass, and Herman Melville would all be invited to look at the stars through the same telescope that William had used when he introduced Maria to astronomy.

When Samuel Haynes Jenks described William's astronomy lecture as "remarkably happy" he captured the sense of joy and wonder that always characterized William's astronomical conversation. Many years later, Mitchell wrote to his nine-year-old grandson:

> Twenty days have passed since the date of thy very nice and acceptable letter. Twenty times this old earth on which we live has turned completely over on its imagined axis. In that period by this motion of the earth only, thou hast moved more than three hundred and eight thousand miles. But this is not thy greatest journey. By the motion of the earth toward the point at thy right hand when thou stands facing the sun, thou hast moved thirty one millions of miles in the same period. All this is perfectly true though we do not perceive it. It is not manifest to the senses so directly as a journey to Boston; but it is as manifest to the understanding.[11]

William was delighted by astronomy; he shared his sense of joy and playful wonder with his daughter Maria, and with all of the girls, boys, women, and men who attended his lectures or climbed the steep stairs to peer through his telescope.

But although William could teach her how to identify the stars, Maria had to learn higher mathematics by herself. From age twelve to fifteen, she studied advanced mathematics with Cyrus Pierce, who briefly led a school on the island. After that, she was on her own, though she always seemed to be able to get her hands on good mathematical texts. At least she had access to her relative Walter Folger's extensive mathematical library and her father's state-of-the-art equipment. Phebe Mitchell Kendall, Maria's beloved sister, commented, "At the time when Maria Mitchell showed a decided taste for the study of astronomy there was no school in the world where she could be taught higher mathematics and astronomy. Harvard College, at that time, had no telescope better than the one which her father was using."[12]

In the 1820s, the Mitchells were not wealthy, but William Mitchell was one of the most respected men on the island, perhaps in the state. He was best known as a teacher but he changed jobs frequently, as many men on Nantucket were forced to do if they wanted to avoid shipping out. If a whale ship came in heavily laden and there was a sudden need for barrels, he might turn cooper for a few weeks; if someone with a good sense of practical chemistry was needed to turn whale oil into soap, he would try his hand at soap making. In his brief memoir, he laughingly listed all of the jobs he had tried in his ceaseless struggle to make ends meet:

> A cooper, a soap boiler, an oil and candle manufacturer, a farmer, a schoolmaster, an Insurance Broker, a Surveyor, a Chronometer rater, an astronomical observer for the Coast Survey, Justice of the Peace, Executor of Wills and Administrator of Estates. Also writer of Wills, deeds and other instruments, Cashier of a bank, treasurer of a Savings Bank. And, without emolument, a member and for some years President of the board of Trustees of the Nantucket Atheneum—a member and for some years Chairman of the Board of Admiral Coffin's School. For many years I was the Chairman of the Committee for the Superintendence of the Observatory of Harvard College. At two different periods, I was clerk of the Nantucket Monthly Meeting of Friends, once for 10 and once for 5 years.[13]

Although many of these jobs may sound menial, and the family's economics were hand to mouth for many years, the provisional nature of Mitchell's jobs was not demeaning. Whenever there was a position of responsibility to fill, William Mitchell was one of the first people to be called on. He served briefly as a state legislator and worked for the U.S. Coastal Survey as one of the first professional astronomers in the United States. They didn't pay him much, but they did provide him with the latest instruments and a sense of connection to the larger—though still very small—community of American astronomers. For the Mitchells, the respect of their neighbors was better than wealth, and on an island that valued intellect as much as Nantucket did, their social place was assured.

By 1845, William Mitchell was an astronomer of national reputation, and when the United States Naval Observatory was opened that year in Washington, D.C., Harvard's Benjamin Pierce proposed William for the directorship. Pierce might have been aware that William's sole astronomical weakness was that he was not a trained mathematician, but Pierce also knew that he worked very closely with one. Pierce ended his letter: "I cannot conclude my recommendation without alluding to the aid which Mr. Mitchell will undoubtedly receive, in his labours, from an accomplished mathematician and excellent observer in the person of his daughter."[14]

William loved the life he made for himself; it allowed him to spend most of his time with his family, home in their crowded and cheerful parlor, rating the chronometers of the captains about to ship out on voyages of three, four, or even five years. A chronometer was a very accurate portable clock, and if it was carefully rated, so that it showed the exact moment when it was noon at the Nantucket meridian, a captain could determine his longitudinal meridian on any given day by comparing the difference between the actual noon on the deck of his ship and the Nantucket noon ticking steadily along in the brass casing of his chronometer. William Mitchell himself traveled very little, but the chronometers that he calibrated traveled all over the globe—and his neighbors trusted their lives to his astronomical expertise. He wrote, "I have no cause to regret that I have given to the objects of the firmament something more than a mere gaze, although, like the poet's muse, Astronomy 'found me poor at first, kept me so.' It is wealth to me, however, to look back upon the astronomical events and phenomena which it has been my privilege to witness a little in the light of science."

Maria Mitchell was an exceptional woman, but she was not as much of an exception to the rule as we might expect. Research by historians Miriam Levin and Kim Tolley has shown that in the first half of the nineteenth century, girls and women in the United States were more likely to be taught science than boys and men were. Boys tended to focus on the classics, which might lead to studying the law and entry into public debates. In contrast, knowledge of the regular movement of the stars or of Linnaeus's biological kingdoms reinforced girls' understanding of the world as an orderly, hierarchical place. No one, it seemed, would become a revolutionary by studying science, which at the time described the world as a stable natural reflection of God's divine hierarchies. In the social hierarchy, of course, women were at the bottom. Studying nature's hierarchies could only keep them in their place.

Even in the context of a time when girls were steered toward science, the situation on Nantucket was unique. The Quakers who dominated the island had a long tradition of female education, as it is a central tenet of the faith that women's minds and souls are as important as men's. Maria Mitchell's brother Henry thought the culture of Nantucket had been profoundly shaped by the Quakers' emphasis on women's education. In a memoir about his famous sister, he explained that Quaker "Discipline," as the behavioral guidelines were known, "as far back as 1695 included the equal education of both sexes."[15] Friends were encouraged to instruct their children in French, German, Dutch, and Danish, but Henry Mitchell relates that by the 1730s, Nantucketers had decided that these languages were somewhat impractical. Instead, island children of both sexes learned "rudiments of the sciences, especially those related to navigation."[16] Henry Mitchell described the Nantucket of his and Maria's childhood as an isolated community with a collective atmosphere of "intellectual excitement" about science.[17]

Few American towns have been both as isolated and as cosmopolitan as Nantucket was in the late eighteenth and early nineteenth centuries. Many small, tight-knit communities were hostile to new ideas, but Nantucket brimmed with enthusiasm for intellectual culture. It had no institution of higher learning; rather, scientific and mathematical activities were organic, informal, casually domesticated. When she was asked as an adult how she first became an astronomer, Maria Mitchell explained, "It was, in the first place, a love of mathematics, seconded by my sympathy with my father's love for astronomical observation. But the spirit of the place also had much to do with the early bent of my mind in this direction. In Nantucket, people quite generally are in the habit of observing the heavens, and a sextant will be found in almost every house."[18] Knowledge of astronomy was useful for navigation, but many Nantucketers' interest in it went far beyond practical application. For some Nantucketers, it was purely entertainment; people wandered around with sextants and talked about planets and comets because there really wasn't much to look at on treeless Nantucket apart from the changing sky. For others, "natural philosophy" offered entrée into international intellectual communities in a way that moral philosophy did not. As mercantile Quakers, Nantucketers had little hope of joining aristocratic or orthodox intellectual circles, but the rapidly growing realm of natural science was far more democratic than any other scholarly field of the time. As all of these unique factors came together in a small seafaring town where there were often far more women around than men, girls and women on Nantucket were encouraged to explore many topics, including scientific and mathematical ones. Perhaps it was only possible for Maria Mitchell to become the American Urania because Nantucket was so firmly ruled by this particular muse.

Benjamin Franklin's mother, Abiah Folger, was a Nantucket Quaker, and she might have passed on to her son some of the local intellectual attitude. At one point, Franklin tutored a young woman named Polly Stevenson, the daughter of his British landlady, in the sciences, loaning her books and exchanging long letters filled with explanations and assignments; they corresponded for more than thirty years. When they started in 1760, he presented the teenager with books that afforded "a good deal of philosophic and practical Knowledge, unembarras'd with the dry Mathematics us'd by more exact Reasoners, but which is apt to discourage young Beginners." He encouraged her to take careful notes, to look up every unfamiliar "term of science," and to ask him about anything she had trouble understanding: "When any Point occurs in which you would be glad to have farther Information than your Book affords you, I beg you would not in the least apprehend that I should think it a Trouble to receive and answer your Questions. It will be a Pleasure, and no Trouble. For tho' I may not be able, out of my own little Stock of

Knowledge to afford you what you require, I can easily direct you to the Books where it may most readily be found."[19] Franklin respected Polly's intellect, and he thought studying science was perfectly appropriate for her; indeed, he once he commented teasingly that after she'd learned all this natural philosophy, everyone would want to marry her: "But why will you, by the Cultivation of your Mind, make yourself still more amiable, and a more desirable Companion for a Man of Understanding, when you are determin'd, as I hear, to live Single?"[20] It is striking that a scientist and statesman of Franklin's reputation would devote so much time to the careful nurturing of one girl's interest in science, but it is even more striking that neither Franklin nor Stevenson found her interest unfeminine, and in fact they might have seen it as socially desirable. Stevenson would eventually marry William Hewson and disappoint Franklin, who, in spite of teasing her about being single, had dearly hoped she would choose his son as her husband.

Walter Folger was the Benjamin Franklin of Nantucket. Said to look exactly like his more famous second cousin, he also shared Franklin's talents, albeit on a smaller scale. Folger made astronomical discoveries and published his findings in the *Journal of the Royal Society,* of which his eminent cousin was one of the few American members. Folger was also a legislator and an inventor best known for his beautiful astronomical clock, which is still on display at Nantucket's Whaling Museum. Like Franklin, he thought science a perfectly appropriate subject for girls to study; in the 1790s, he did not hesitate for a moment to teach his younger sister Phebe Folger algebra, calculus, and navigation, and in the 1820s he was founding president of the Nantucket Philosophical Institute, which welcomed women as members. Walter and Phebe spent so much time studying together that at one point their mother, Elisabeth Starbuck Folger, remarked, "Walter has his sister Phebe in that Algebra class he's teaching—he'd better teach Anna [his wife]!"[21] And so Phebe Folger was carefully educated in art and language, science and mathematics. The Quaker community on Nantucket appreciated her skills: in 1796, she was appointed mistress of the Quaker school.[22] In 1797, she started work on an ambitious commonplace book, a leather-bound book of manuscripts, paintings, and equations. The title page proclaimed that her notebook was "Un Recueil" that would include "Painting, Penmanship, Algebra, and Pieces selected from various authors in prose and verse, with a few pieces in French with their translation by Phebe Folger of Nantucket."[23] She continued to add to it for over fifty years. The notebook is as marvelous as Walter's astronomical clock. Many of the beautiful watercolors are copied from drawings in magazines or books, and many of the literary extracts are also copied (in overwhelmingly beautiful handwriting). But it also includes much original work. Phebe made several paintings of the Nantucket landscape (which are still familiar to Nantucketers), and some of the poems are Folger's own, as are many of the mathematical problems and the translations of French literature.

One of Folger's poems, written in 1809 when she left Nantucket to start a farm in New York State, describes the educational encouragement and support she received from the Nantucket community, and particularly from her family. Folger bade farewell to:

> ...you with whom my youthful days were past,
> Days which we might with reason, wish to last,
> With whom I've turned a philosophic page,
> Or traced the manners of each distant age,
> Or sought the realms of science to explore,
> Or cull'd from fancy's richest work her lore,
> Oft will kind memory bring those days to view
> And fancy's pencil paint the scene anew.[24]

Poems by other Nantucket women echo and amplify Phebe Folger's celebration of educational opportunity. Written in 1815, Lydia Wing Coleman's "Farewell Lines" might respond to Phebe Folger's "Farewell" directly. Lydia Wing Coleman was a distant cousin of Maria's mother, Lydia Coleman Mitchell. She wrote her poem at the end of a long visit to New York, just before she set sail for her Nantucket home. Coleman was returning to the island that Folger was leaving behind, and her goal in the poem was to encourage a young female friend to continue to pursue her intellectual, and specifically scientific, interests:

> Now duty summons me away,
> My sand-girt Isle allures me too:
> Yet, yet forgive my lingering stay,
> To bid the friend beloved, adieu.

Farewell to thee, my youthful friend;
May heavenly wisdom gild thy days;
May guardian angels thee defend,
And lead thee through life's dangerous maze.

May science fair thy brows entwine;
May thou be rich in wisdom's lore;
May thou fame's rugged ascent climb,
And every classic path explore.

May genius weave a garland fair,
And cull her pearls from learning's stem,
And thee select, the same to wear,
As fittest for her diadem.

Coleman's remarkable poem offers clear evidence that Nantucket women encouraged one another to study, and particularly to study science. The poem is conventional in its prosody and rhyme scheme, sentimental in its tone; it is not intended to be a radical document. But it does reveal an attitude toward girls' intellectual ambition that seems radical now. Coleman's blessing on her youthful friend is genius. She urges her to aspire to fame, to climb the mountain of wisdom, and to produce pearls of wisdom that will adorn genius's crown. And the first among all of these blessings is clear: "May science fair thy brows entwine." "Farewell Lines" must have been well known and respected in Nantucket, because it was carefully preserved and, in 1853, more than twenty years after the poet had died, it was published with other poems in a small volume entitled *Seaweed from the Shores of Nantucket*. In Nantucket Historical Association records and in the volume itself, Lydia Wing Coleman is referred to as "The Poet."

Yet another scientific Nantucket girl, Mary S. Coffin, was known as a poet who "specialized in writing verse about Franklin, Newton, Bacon, and other contributors to science."[25] As the following excerpt from her 1852 poem "Education" shows, Coffin thought of her poetry itself as an offering to the scientific community and hoped that she would make an enduring contribution to science:

Like the fixed stars from their far distant height
Learning reflects a pure and heavenly light,
By which the everlasting ages shine

And bless its holy rays with joys divine,
Ah! Would to heaven the exalted gift were mine,
To lay, fair Science, on thy sacred shrine
A worthy offering, fitted to inspire
The ardent breast with an increased desire
To sound thy mysteries, improve the mind
And strive to elevate and bless mankind!

It is tempting to trace a trajectory of growing intellectual confidence that moves from Phebe Folger's gratitude for her brother's encouragement to Lydia Wing Coleman's encouragement of another young girl and then to Mary Coffin's open avowal of her own ambition. However, the slender thread of an indefinite chronology may not support this narrative. Rather than reading for progress, it is probably more realistic to imagine all of these activities as simultaneous: girls were encouraged to study science by family and neighbors alike. They encouraged each other, and they found great fulfillment and a sense of community in their studies. They believed it was possible for each one of them to make real contributions.

This atmosphere of shared scientific interest and excitement, this welcoming and encouraging coterie of scientific women, was a key factor in Maria Mitchell's success as a scientist. By the time she was ten or eleven, Maria knew that it was entirely natural for a girl to hope for scientific glory, and when she was twelve and a half, she had her first chance to do important scientific work.

An annular eclipse was expected on February 19, 1831, and Nantucket was the only location in America from which it could be entirely observed. In preparation, William took panes from the parlor window and set up his Dolland telescope in the front room. He explained every step to Maria, and they set the telescope carefully and waited. When the shadow came, William watched through the smoked lens of his telescope, and Maria steadily counted the seconds. They recorded their observations together, and Maria treasured their notes for the rest of her life. It was the first time she had helped to do original astronomical research; no one else in the world had had the chance to make the observations she and her father had made. Because they had coordinated their observations with those of astronomers in Monomoy, on Cape Cod, and in Dorchester, south of Boston, the Mitchells were later able to calculate the exact longitude of their front room on Vestal Street. From then on, when the Mitchells rated a chronometer, it would be absolutely accurate—and perhaps more important, Maria Mitchell would consider herself a true astronomer. Soon afterward, Maria Mitchell rated her first chronometer on her own; this was minor—a commercial venture, unlike the eclipse observation, which was pure science. But it was for rating the chronometer that she earned her first payment for astronomical work. At thirteen, she was well on her way to becoming a professional astronomer.

Like most young astronomers of the day, Maria acted as assistant to a beloved family member, but that did not mean that she was not an astronomer in her own right. In the same year that Mitchell observed the eclipse with her father, the Nantucket Philosophical Institution, the island's version of a Royal Society, decided to admit women. On December 1, 1831, the minutes record that "the usefulness of the institute will be extended, and interests of the community benefited by such a modification of our constitutional by-laws as will afford a more free access to our association than has hitherto been presented.... It is therefore recommended that females be admitted free of charge as members of the association."[26] At the very next meeting, more than a score of women joined the Nantucket Philosophical Institution. Walter Folger was president of the group and William Mitchell was vice president, so it should come as no surprise to read that several Mitchells were listed as new members: Lydia C., Sally, Maria, Sarah, and Mary. That same day, there was a motion to admit Phebe Folger Coleman, who had bid a fond farewell to Nantucket's intellectual community in 1809.

In some respects, the Nantucket Philosophical Institution was typical of the early institutionalization of science, the efforts to create a scientific community that led to the British Royal Society and the Institut de France. But Nantucket was atypical in that it admitted so many women at the start. Although the Royal Society admitted two women in 1835, most European scientific institutions refused to admit any women at all. In contrast, by 1832 there were as many women as men in the Nantucket Philosophical Institution.

Of course, being a member of the Nantucket Philosophical Institution was nothing like being a fellow of the Royal Society. Membership did not signify scientific greatness but merely an interest in scientific matters. Even so, when Maria Mitchell signed her name in the institution's log alongside those of her mother and sisters, joining a group to which her father and her well-known scientific cousin Walter Folger were proud to belong, it signified a great deal. Maria had the good fortune to grow up on Urania's island, where a girl who wanted to be an astronomer had almost limitless opportunities.

Nantucket Athena

By the early nineteenth century, education in the United States had become a public affair. Almost every town provided free public schools for young children, and many were in the process of building high schools. Inspired by the ethic of self-improvement that had been fostered by notables such as Benjamin Franklin, Americans were forming and joining hundreds of intellectual clubs and associations. Growing out of these associations, the athenaeum movement took hold: intellectual towns across New England and the mid-Atlantic built grand libraries and art galleries known as athenaeums, which became cultural hubs.

Although Mitchell's experience at home—her education in the parlor, if you will—was shaped by English writer Maria Edgeworth's tales, Mitchell's childhood experiences on Nantucket were very different from Edgeworth's isolated childhood in suburban London and colonial Ireland. Surprisingly, there were far more opportunities for learning on the small Massachusetts island than there were in Britain or Ireland. A wealth of schools, clubs, and libraries awaited the Mitchells beyond their front parlor, and all were open to girls. The situation was much different in England, where female education was the province of privately hired governesses who lived in their pupils' homes. School was for poor English girls, and few of them had the chance to attend.

But the burgeoning wealth of educational opportunities outside the home was very new in Nantucket, and even there, the domestic

education that children gained in their parlors was consciously struc-
tured as an important social good. Historians such as Linda Kerber
and Margaret Rossiter have explored the importance of the home
education provided by the doughty mothers of the young republic,
and historian Sally Kohlstedt points out that many fathers were also
involved in the educational work of the parlor. Scientific education
in particular took place at home more than anywhere else.[1] To put it
even more bluntly, in the late eighteenth century and the early nineteenth century, before the first scientific associations formed, there
really was nowhere to study science outside the home. Eighteenth-
century schools focused on reading, writing, and arithmetic, and
colleges on classical languages and theology. Scientific paraphernalia
lia—microscopes, telescopes, and cabinets full of rocks, feathers, and
seashells—were furnishings for middle-class parlors long before they
were allowed into the classrooms.

The parlors of Nantucket were crowded with such curiosities.
Scrimshawed whale teeth, carved Polynesian masks, and Chinese
lacquered boxes sat next to birds' nests, scallop shells, fossils, and the
ubiquitous sextants and spyglasses. Globes, maps, and charts were
everywhere, and many families owned a painting or two showing a
ship in a faraway Asian or Pacific harbor.

A few blocks from the Mitchells' house was another, far more
stately house that also played an important role in Maria's education:
Walter Folger's house on Liberty Street. Unlike the Vestal Street
house, which was small and chaotic, Folger's house was large and
quiet. His wife was timid, and his children were grown. Walter's
front room was full of books, including all of the latest mathematical
and scientific treatises. He had published two articles on astronomy
in the *Memoirs of the American Academy of Arts and Sciences,* and piled
on his shelves were copies of many of the earliest scientific journals.
A year after Maria was born, in 1819, Walter Folger began constructing a Gregorian reflecting telescope that had a five-inch glass.
When he wasn't in the parlor reading or making notes for one of his
lectures to the townspeople, or writing an article on astronomy for
the *Nantucket Inquirer,* he was often in his workshop, figuring out
how to build what would become one of the largest telescopes in the
United States.

Although the telescope intrigued all of the Mitchells, the most
impressive piece in Walter Folger's house was the clock he had built

around the time his cousin William Mitchell was born. Many Nantucket parlors gave pride of place to handsome grandfather clocks; the Mitchells had one of their own. But Folger's clock was unlike any other. It was built from the works of an old grandfather clock, to which Folger added carefully milled orbits, levers, and weights. He made a disk of gold to represent the sun and then milled a perfect sphere of silver, half enameled in black, for a moon. He built an ecliptic so his sun could change its angle to the horizon. Next, he added the date, including year and day. He crafted the zodiac, carefully figuring in the tides at 'Sconset and the seasonal changes in the moon. The works for his moon revolved once every eighteen years.

Walter Folger's astronomical clock is a beautiful thing, and Maria studied it every time she went to the Folgers' house. From the outside the clock looks simple, elegant, and inevitable, but its works are remarkably intricate. The sun rises on the clock face every morning at precisely the time it rises in Nantucket: one morning at, say, 5:43, the next morning a few minutes earlier or later. It is perfectly predictable—a remarkable response to the French Revolution. Folger built it as the French Revolution raged, alternately studying French in order to read the latest news and feverishly diagramming the revolving works for his clocks. Revolutions among humans can be bloody and terrifying, but the celestial revolutions of planet Earth are calm, grand, and beautiful.

When Folger was building his clock in the late eighteenth century, astronomers saw the universe as something very similar to a clock. Newton's theory of gravitation explained the works; as the gears moved each hand around the face of a chronometer, so the strict mechanical laws of gravitation pulled each planet along its elliptical orbit. Folger himself explained:

> What subject of human contemplation shall compare in grandeur with that which demonstrates the trajectories, the periods, the distances, the dimensions, the velocities and gravitation of the planetary system; states tides, adjusts the mutations of the earth and contemplates the invisible comet wandering in its parabolic orb.... Language sinks beneath contemplation so exalted and so well calculated to inspire the most awful sentiments of the great Artificer, of that Wisdom which could contrive the stupendous fabric; of that Providence which can support it, and of that Power whose hands could launch into their orbits—bodies of a magnitude so prodigious.[2]

Walter Folger was a second cousin of Benjamin Franklin, and he was the same sort of deeply conservative revolutionary. He didn't hesitate for a moment to study French with his little sister during the bloody French Revolution, nor, later, did he hesitate to teach her algebra, calculus, or navigation, at which time he wrote that she was "ingenious, inventive, imitative and mathematical."[3] To us, "imitative" seems out of place in this list of attributes, but at the close of the eighteenth century it was high praise. Walter Folger's inventiveness was not the same thing as originality—he didn't invent sunrise, the full moon, or the geometric regularity of the relation between the moon's cycles and the tides on Nantucket's shores—but he did figure out how to imitate them in carefully milled metal. Similarly, Walter and his community trusted that an intelligent girl who studied mathematics and carefully observed nature would be imitative in the best sense of the word: she would conform to the grand and benign rationality of nature itself. In a clockwork universe, studying the revolutions of the heavens was far safer than studying human revolutions; mathematics (such as Newton's) led to a certain knowledge of everything's proper place in the orderly universe.

Walter Folger inspired many young Nantucketers, including girls, with his scientific achievements. Maria Mitchell was fairly close to her cousin, but he was a remote figure, called "odd as huckleberry chowder" by some islanders.[4] She and her father often discussed astronomical problems with Folger, who had been at least as prominent among the astronomers of his earlier generation as William Mitchell was among his.

In the 1820s, American science began to move out of the parlor. Walter Folger and William Mitchell were among the first in the United States to form a scientific association when they founded the Nantucket Philosophical Institution in 1826, but they were never terribly close to each other. Their worldviews were too different: Folger's clockwork astronomy was alien to the Mitchells, who had been shaped by the tenets of Romanticism. Walter Folger was a creature of the eighteenth century, William Mitchell of the nineteenth. Walter Folger was from a generation that had little access to, or interest in, formal education. During his childhood there had been no established schools on Nantucket, and the few colleges that existed in the United States were generally closed to Quakers. Now, all of this was changing. William's developing interest in collaborative communities of scientists and students distinguished him from his older cousin, who did his work in isolation. Folger was the figurehead of the Nantucket Philosophical Institution, but William Mitchell, who served as its second president, was its guiding spirit.

The implications of the new commitment to public education were enormous; the inclusiveness of the American public school movement set the stage for the broad expansion of rights (to working-class people who did not own property, to people of color, and to women) that made the nineteenth century such an optimistic and exciting time. The spread of scientific education in the United States was closely related to the varied student body of the common schools, since scientific subjects were considered safe and appropriate for women and the lower classes. Sally Kohlstedt explained that "social studies of science remind us that the status and authority of various sciences are intimately connected to the . . . institutions that express and define them." In early-nineteenth-century America, unlike anywhere else, young women were hired by the hundreds to teach in elementary schools, which were administered by town governments but considered part of the feminine domestic sphere. And so American public schools focused on the sorts of scientific education that had thrived in eighteenth-century families.

The historian of education Carl Kaestle calls the 1830s the start of "the feminization" of the American common school classroom.[5] A look at Nantucket's common-school debates will explain one of the reasons that school was feminized: economics. The bald reality was that it was significantly cheaper to hire women teachers than men teachers. Because of the general presumption that women were not responsible for housing or feeding themselves, a woman's salary could be small; it was just a bit of "extra money." This argument carried the day on Nantucket as elsewhere, and if there had not been a supply of well-educated girls willing to work for low wages, Nantucket might never have opened a public school.

In 1645, Massachusetts passed a law requiring every community to provide a school for every group of fifty children. In spite of their commitment to education, the Quakers of Nantucket "did not take kindly to the idea of free public schools" that would admit non-

Friends, according to the Nantucket historian Alexander Starbuck.[6] For about two hundred years, Nantucket simply refused to comply with the law. In 1817, Samuel Haynes Jenks, who had been born and raised in Boston, complained to the town meeting, but they laughed at him, calling him "a stranger and a coof" (Nantucket slang for an off-islander). He sued the town, and in 1818, the year Maria Mitchell was born, the town appointed a common-school committee to do some research. The committee reported that "school-mistresses of good moral character and well-qualified to teach youth from three to twelve years old may be procured to teach a school of fifty scholars for one hundred and four dollars a year."[7] Working with a budget of a thousand dollars, the school committee figured that they could set up four female schools—that is, led by school-mistresses— and still have enough money for one school taught by a man. Doing the math, they figured out what would be left to offer the male teacher: "We are also of the opinion that one man school ought to be supported by the town, the annual expense of which we calculate at Three hundred and Eighty four dollars—so that five schools, four to be taught by women and one by a man, may be maintained during the year for one thousand dollars."[8] The salary they projected for the man ($384) was more than three times that projected for each woman ($104). Women had few other options for employment; factory work would not become available for decades. Men, on the other hand, could earn more money doing almost anything else, and many occupations were less arduous than teaching. Across New England, similar economic calculations quickly led to the overall feminization of school.

Nantucket dragged its feet for ten more years, but in 1827 the island finally established a public school. The first principal was William Mitchell, a reliable and much-loved educator who advocated the new, gentle methods of pedagogy that women teachers were bringing into the classroom. Mitchell ran the school for three years, from 1827 to 1830, but then left to start his own private school on Howard Street, "a select school for fifty scholars, half of each sex."[9] He ran this school from 1830 to 1833.

There were very few public high schools at this time, but the few that were established were similar to the private academies that were springing up everywhere. Most were single-sex, and the curriculum was very different for girls than for boys, who generally avoided secondary education unless they were aiming for college. Girls' secondary schools, like elementary schools, often employed women as teachers and tended to embrace the scientific subjects that had first thrived in middle-class parlors. Boys' secondary schools were presided over by men and tended to emphasize the classics.

Maria Mitchell attended her father's schools from age nine to fifteen, and her educational philosophy was shaped by the experience. William Mitchell was frankly nonauthoritarian. In his schoolroom, he said, "Punishment was almost unknown. We met together as common friends and for mutual improvement."[10] Maria later attended Cyrus Pierce's private academy, the most advanced educational opportunity on the small island, but at fifteen, she had learned all that the available schools had to teach her, and she got her first job as a teaching assistant to Cyrus Pierce. Pierce was a fabulous teaching mentor: when Nantucket established a high school in 1837, he was appointed its first principal, and he served in that capacity until Horace Mann appointed him to be the founding director of the world's first school of education, the Framingham Normal School, located outside of Boston. Now called Framingham State College, it was founded for the express purpose of training teachers to teach. After three years of assisting Pierce, Mitchell was as well educated as it was possible for a girl to be in 1835 in Massachusetts. She might conceivably have studied for a few more years if she had been able to attend Emma Willard's seminary in Troy, New York, and, of course, if she had been a boy, there would have been the option of Harvard or one of the new colleges that were opening across the country. But the first experiments in the higher education of girls were a few years in the future, and it is doubtful that even these could have given Mitchell the level of education she gleaned from her family, her community schools, and the clubs, associations, and institutions that were open to her on Nantucket. Thanks to her father's prominence, Maria befriended George Bond, who was working as an assistant to his father, William Bond, in the fledgling observatory at Harvard; she met Benjamin Pierce, the great astronomical mathematician; and she corresponded with the director of the United States Coast Survey, Alexander Dalles Bache, one of her father's many employers. Her resources were well-nigh incomparable.

Moreover, college was not the place to learn science. American colleges, like the British institutions on which they were modeled, were founded for the primary purpose of preparing men for careers in the church, so in the early nineteenth century, relatively few men attended them. By the 1830s, college curricula had begun to expand, and colleges were seen as good places to acquire social polish, which could be a great asset in a number of careers. But many professions—law and medicine, for instance—still followed apprenticeship models rather than requiring degrees as credentials; and science was not yet a profession. Among men of science, neither Benjamin Franklin nor his Nantucket cousins Walter Folger and William Mitchell had attended college. As a Quaker—and an enlightened skeptic—Franklin would have been entirely out of place at puritanical Harvard, at Yale (whose graduate Jonathan Edwards had penned the sermon that condemned the "Sinners in the Hands of an Angry God"), or at the even more religiously conservative Amherst College. By the time William Mitchell was old enough to apply, Harvard had somewhat liberalized, but it still offered little in the way of a scientific education. A better place to learn science was Nantucket, and Maria began to contribute in kind when, at the age of seventeen, she founded her own little school.

The advertisement ran in the *Nantucket Inquirer:*

SCHOOL

Maria Mitchell proposes to open a school for girls on the first of next month at the Franklin School house.

Instruction will be given in Reading, Spelling, Geography, Grammar, History, Natural Philosophy, Arithmetic, Geometry and Algebra.

Terms $3 per quarter. None admitted under six years of age.[11]

On the school's first day, a sunny September morning in 1835, seventeen-year-old Maria Mitchell left home early and walked along the quiet, sandy lanes to Franklin Hall, the community center building on Trader's Lane. She had rented a room and arranged the tables and chairs for class in advance. She had borrowed as many curiosities from home as her parents would let her carry; now all she needed was a student or two.

As she waited, Mitchell had no idea what to expect. She hoped she would have enough students to cover the cost of renting and heating the room and that she would earn a bit more money for herself than she'd earned as Cyrus Pierce's assistant teacher. The first children to enter were three dark-skinned "Portuguese" girls, probably of Cape Verdean extraction. Most nonwhite townspeople on Nantucket at the time were quartered with the Cape Verdean community, and generally referred to as "Portuguese." One little girl spoke up to ask if they could enroll, and Mitchell decided on the spot: she couldn't look into the young girl's bright eyes and deny her. Community conflict about integrating Nantucket's public common school had raged the previous winter, and Mitchell knew that her snap decision to allow students of color in her fledgling school might alienate parents and potential patrons, but Mitchell believed that every girl should have the chance to learn, and she had never lacked courage.

And what a chance it was! Maria Mitchell's school must have been an exciting place. In one large room, Mitchell had students ranging in age from six to thirteen or fourteen. Like most classrooms at the time, hers was probably chaotic by today's standards; clusters of students of different ages worked simultaneously on subjects ranging from basic reading to Shakespeare, arithmetic to spherical geometry. Mitchell took them outside as often as she could, meeting them late at night to show them the stars through her father's telescope, or early in the morning to collect plants and water samples in the bogs.

The school was a success, but it only lasted a year. In 1836, having established a professional reputation as a doggedly persistent worker and a cheerful, inspiring teacher, Maria Mitchell was offered one of the most prestigious jobs on the island: the librarianship of the Nantucket Atheneum.

The Nantucketers decided to spell the name of their athenaeum a little differently than others, but they were part of the athenaeum movement that swept across the major cities and towns of England and America in the early nineteenth century. The first athenaeum opened in Liverpool in 1798, its name meant to indicate that it would be a temple "dedicated to Pallas Athena, goddess of wisdom and learning and of the useful and fine Arts." In 1807 the Boston Athenaeum followed, aiming to "principally be useful as a source of information and as a means of intellectual improvement and pleasure. It is a fountain at which all who choose may gratify their thirst for knowledge."[12]

As the name implies, athenaeums were not mere libraries; they were temples of wisdom, conceived and designed as places to meet and talk about ideas. They were not public—people bought shares in the organization—but they were not necessarily exclusive. Each athenaeum had a different policy about who was allowed in the door. Philadelphia's, for instance, did not allow women even to enter the building and discouraged nonmembers from visiting unless they were the invited guests of members. Needless to say, Philadelphia's athenaeum quickly took on the character of a private club. Boston's was a bit less exclusive; the Boston Athenaeum invited Lydia Maria Child to join in 1824 when she published her first novel (*Hobomok*), but later changed its collective mind and ejected her when she publicly advocated interracial marriage. Still, the Boston Athenaeum was a relatively welcoming place. Women wandered the galleries and drank tea together in the tearoom.

At the other end of the spectrum, the Nantucket Atheneum outright opposed exclusivity. Perhaps this was because it rose out of the Nantucket Philosophical Institution. With the exception of Walter Folger, most of the directors of the NPI were also on the board of the fledgling athenaeum. The proprietors' review made their inclusiveness explicit in the organization's 1853 annual report, concluding: "The difference between a community well-supplied with books, and those generally available to all classes, and a people poorly supplied and this supply in the hands of a few must be enormous—none can estimate it. May we hope then for a general desire to sustain the Nantucket Atheneum, a general willingness to devote a little money yearly to increase the Library: We feel assured that we may."[13]

For David Joy, one of the two founding patrons of the Nantucket Atheneum, this mission emphatically included people of every description—rich and poor, white and dark-skinned, male and female. Joy was a friend of Frederick Douglass and many other well-known American reformers. When he died in 1875, the radical reformer Adin Ballou commented that Joy was not only a friend to African Americans, but also "ardently wished to see women elevated by education, spheral opportunity, legal prescription and personal exer-

tion."[14] The great abolitionist William Lloyd Garrison also celebrated Joy's commitment to expanding rights, asking "in what well
ordered movement for the co-equal education and enfranchisement
of our race, irrespective of nativity, complexion, or sex, did he not
take an interest?"[15] Seth Hunt, a freethinking abolitionist, remarked,
"He gave his zealous sympathy and aid to the work of breaking down
the partition walls of caste, set up by prejudice and bigotry, for his
great soul burned with indignation at the unjust and cruel policy
which makes sex or color a ground of obloquy and a reason for with-
holding the rights and privileges of civil government and other institutions."[16] For David Joy, the Nantucket Atheneum was part of a
larger movement that offered equal education to everyone.

In the context of the 1830s, when men and women were beginning to be sorted into the separate spheres of the public and the domestic, Joy's commitment to "spheral opportunity" was significant.
School was a private, domestic, feminine place, but it was also far
more public than it had ever been. The Nantucket Atheneum embodied a fascinating elasticity of spheres; it was both public and private, an institution that was not a university but invited all
individuals to make their own personal efforts toward intellectual
self-culture. It had no professors or teachers, no top-down model of
learning; it was designed as a haven for autodidacts. From the start,
the Nantucket Atheneum offered a generous schedule of lectures.

The job of supervising the Atheneum's library was a plum. Why
was it offered to an eighteen-year-old schoolmistress with only one
year of teaching under her belt? The founders might have decided to
hire a woman because their budget was limited or because there was
a shortage of male workers on the island due to the whaling trade.
However, they probably chose Mitchell specifically because she was
radiantly intelligent, ferociously witty, and demonstrably knowl-
edgeable about "Reading, Spelling, Geography, Grammar, History,
Natural Philosophy, Arithmetic, Geometry and Algebra," not to
mention literature and astronomy.

The year 1836 was good for the Mitchells all around: William
was offered the directorship of the Pacific Bank, which was housed
in a grand brick building at the top of Main Street. The job paid a
generous salary of $1,200 a year plus tenancy of the elegant apart-
ment above the bank. The Mitchells moved to Main Street, and for

the first time there was plenty of space for everyone in the family. The difference between the Mitchells' new high-ceilinged parlor, with its large windows looking down toward Straight Wharf, and the small and crowded kitchen–cum–sitting room on Vestal Street was emblematic of the boom in the town's fortunes in the 1830s and 1840s. The family quickly filled the parlor with scientific curiosities, and one day, when both William and Lydia were away, the girls smuggled in a piano. The stringent Quaker Discipline forbade musical instruments, but William and Lydia did not have the heart to impose such strictures on their children. The bank became a lively, often joyful hub of social activity.

Later that year, the well-known astronomers Elias Loomis, from Yale, and William and George Bond, the father-and-son team who ran the Harvard Observatory, came to Nantucket to help the Mitchells build a small observatory on the roof of the Pacific Bank. They fixed a ship's mast near the chimney to provide stability for an equatorially mounted telescope that could be precisely aligned with the earth's axis. The roof walk was much larger than the one on Vestal Street, so Maria and her father could work together comfortably. There was also room for guests—visiting astronomers, often, but many others, distinguished and otherwise, would climb out onto the roof of the bank over the next twenty-five years.

Maria Mitchell's life outside the Nantucket Atheneum was rich, but the Atheneum was probably her greatest good fortune over the next twenty years. She was now free to spend most mornings in quiet study in the library, teaching herself the mathematics needed for the new nineteenth-century astronomy. As a whale ship had been for Herman Melville, the Nantucket Atheneum would be her Harvard and her Yale. Although young, she was perfectly suited for the job; her wide range of interests made her the ideal person to select a useful catalog of books. The library was open to patrons for a few hours each day, but Mitchell was often there for many hours before and after. During her tenure, she read a great deal of literature (it was a great age for poetry, and an even better age for the novel) and patiently taught herself pure mathematics, working through all of the recent books of astronomy. The daughter of a woman who was said to have read every book on Nantucket, Maria Mitchell eventually came very close to that ideal: she certainly read every book in the

Nantucket Atheneum. As host of the institution's many public lectures, she built a remarkable network of friendships with influential literary, political, and scientific figures. She taught Ralph Waldo Emerson how to use the telescope on the roof of the bank; she was there when Frederick Douglass gave his first public speech; and she met both Sojourner Truth and Herman Melville when they came to Nantucket.

Perhaps because of her exposure to these great radical thinkers, Mitchell began to question her Quaker beliefs. She never abandoned her Quaker sensibilities; all her life she dressed in simple Quaker-style clothing, used "thee" and "thou" in intimate conversation, and retained a sense of personal humility that was culturally very Quakerish. But in 1843, shortly after her brother Andrew was asked to resign from the Quaker meeting because of his marriage to a non-Quaker, Mitchell made the great step away from her childhood religion. The minutes of the Nantucket Women's Society of Friends for September 28, 1843, recorded that "Maria Mitchell informed us that her mind was not settled on religious subjects and that she had no wish to retain her right of membership . . . it is concluded to disown her."[17] From that point forward, Mitchell attended the Unitarian Second Congregational Church.

Mitchell's new church allowed her to think for herself with much greater latitude than the Society of Friends had. In this respect, it was very similar to the other institutions that fostered Mitchell's intellectual development. From her happily bookish family and her varied selection of schools, sometimes led by her gentle and open-minded father, to the Nantucket Atheneum and the Unitarian church, Mitchell's early life was full of possibilities for intellectual exploration. In her case, these explorations led irresistibly toward science.

It is significant that Maria Mitchell brought science into the Atheneum, rather than the other way around. By appointing Maria Mitchell as librarian, the Nantucket Atheneum made women and science central to their institution. Henry Albers, the astronomer who would one day be one of Mitchell's successors at Vassar and who would collect and edit her papers with loving attention to detail, wrote in 2001, "As her brother Henry pointed out, she in essence acquired a college education through the books at the Atheneum where she taught herself mathematics, including calculus and the most advanced astronomy of her day."[18] Helen Wright, Mitchell's mid-twentieth-century biographer, wrote that Mitchell "worked like one possessed until she understood the formulae in Bridge's *Conic Sections* and Hutton's *Mathematics* and the significance of Bowditch's *Practical Navigator*... she spent agonizing hours over mathematical tomes and reports of learned societies.... She read the works of LaGrange, Laplace, and LeGendre in the French, puzzled over involved Latin passages in the Theoria Motus Corporeum Coelestium of Karl Freidrich Gauss, and at the same time, taught herself German."[19] In addition, Mitchell was still reading widely in astronomy—keeping up with the British astronomers John Herschel and George Airy, and the advances in mathematics. As librarian, Mitchell decided what books to acquire for the Atheneum's collections. She was also available to act as an intellectual guide to patrons, or just to talk to them about whatever interested them, from novels and poetry to astronomy to practical navigation. She helped boys and girls choose appropriate books on a wide range of scientific subjects, and she took care to smooth over the rough patches—to make challenging subjects comprehensible.

Mitchell's salary from the Atheneum was never very high. After a few years, it plateaued at a hundred dollars a year, a little less than a schoolmistress would have made. In addition to an almost limitless supply of books, the Atheneum gave her oceans of time, since it was only open to the public fifteen to twenty hours a week. Mitchell read, studied, and calculated, and was free to work on observations and surveys with her father. In 1838 she helped him to map Nantucket, surveying much of the island with him, and in 1840 the two of them determined the Nantucket meridian and then laid a marble pillar next to the Pacific Bank to mark where it was. In the years before time was standardized around the Greenwich Observatory, Nantucket whalers sailed on Mitchell time rather than Greenwich time, because the Mitchells calibrated the sailors' chronometers—and so, in a navigational sense, the Mitchells' observatory was at the center of the Nantucket whaling world.

Maria's life was not all books and astronomy. Her sisters spent a great deal of time at the technically forbidden piano, and although Maria wasn't musical, she was always ready to listen and encourage. She entertained the neighborhood children with funny, wildly imaginative stories, and she was the quickest hand in the family at making up silly rhymes. In 1844 she was one of the founding members of the Coterie, a literary club that started with twenty-two men and twenty-two women. According to its constitution, "the object of the Society [was]...two-fold, the first, and most important [was] ...the enlargement and improvement of the mind; for this purpose ...[it was] agreed that every member [should]...contribute her or his proportional share of original composition, in order to carry out one important feature of the Society; the giving of pleasure and profit to every other member of the Society."[20] Maria Mitchell faithfully contributed a poem every Monday, although, since the contributions were anonymous, we cannot tell if her themes were astronomical. One member contributed a letter on the "Moon's Coterie," which described a "lunatic's" view of the earth on the night of an eclipse, while another described an imaginary "Tour to Several Planets." Mitchell might have had a hand in those compositions, but one of her few signed poems from the 1840s, when she was in her twenties, is about school rather than astronomy. Here, she praises an early teacher:

> What though I say unto myself,
> The valued stores of glittering pelf,
>   Dame Fortune may deny me,
> The love of books that maiden taught me,
> The blessing which through life it brought me
>   A Rothschild could not buy me.
>
> What though to me it is denied,
> To roam o'er countries far and wide
>   By bard and poet sung,
> Give me but books, in thought I stroll
> To every land from pole to pole
>   And talk in every tongue.
>   . . .
>
> When I would mingle with the sages
> And seek the learning of all ages
>   Not near the crowd I hover
> I find the wisest of the wise
> The learned whom I highest prize
>   Within a volume's cover.
>
> The angels guard her, whose kind teaching
> So many darkened hearts was reaching

In all her maidenhood
And bless for her the spinster race
Who always fill the highest place
In seeking others good.[21]

This poem offers documentary evidence that Mitchell gave much thought to the importance of education in the 1840s. It is also notable that by 1845, when she was in her late twenties, Mitchell was beginning to identify herself as a spinster and to think that the single state was "the highest place in seeking other's good." Another poem from the same era, "How Charming Is Divine Philosophy," reflects on the same topics:

Did you never go home alone, Sarah,
It's nothing so very bad
I've done it a hundred times, Sarah
When there wasn't a man to be had

And I've done it a hundred times more
When I've seen them stand hat in hand,
I've walked alone to the door,
And they've continued to stand.

And it's useless to try berating,
I gave it up long ago.
Besides, there's a compensating,
Kept up in these things, you know.

There's a deal to be learned in a midnight walk
When you take it all alone.
If a gentleman's with you, it's talk, talk, talk.
You've no eyes and mind of your own.

But alone, in dark nights, when clouds have thickened
And you feel a little afraid
Your senses are all too supernaturally quickened
And you feel a little afraid

. . .

You have only to listen and words of cheer
Come down from the upper air
Which unless alone you never would hear
For you'd have no ears to spare

. . .

Ah Sarah, there's much unwritten lore,
Unconned when others are by,
Which if you have no chance to learn before
You'll learn when you're old as I.[22]

Both of these poems focus on intellectual attainment and independent thinking as well as solitude; in fact, "divine philosophy" seems to be the "compensation" for going unpartnered. It is also worth noting that at this early time in her career, Maria devoted almost as much time to literature as she did to science, reading a great deal of fiction and poetry and producing a regular stream of tales for children and poems for her Coterie. Today, the sciences and humanities are often seen as separate lines of inquiry requiring different skills; indeed, we tend to believe they require different mind-sets, different personalities. C. P. Snow, the English physicist and novelist, lamented in 1959 that science and the humanities had become "two cultures." But until the late nineteenth century, the worlds of human thought were generally seen as contiguous and overlapping.[23] It would be anachronistic to study early- or mid-nineteenth-century science separately from that time period's art. Science was considered a branch of philosophy, and mathematics was the cornerstone of the liberal arts. Science and math *were* philosophy and art; both were understood as highly imaginative activities. The experimental scientific method was already in use, but the emphasis was not so much on the objectivity of the observer as it was on the creative imagination of the person who could design experiments and apparatuses, and foresee results.

Poets turned to science for inspiration, and even for language and metaphors. Often, poets and scientists were one and the same: writers fiddled with microscopes and telescopes while scientists read, recited, and even wrote poetry. Maria Mitchell wrote poems all the time, though she was never very happy with the results. Perhaps the most successful poet among the great nineteenth-century scientists was the physicist James Clerk Maxwell. One representative verse from his "Paradoxical Ode" runs:

But when thy Science lifts her pinions
In Speculation's wild dominions,
We treasure every dictum thou emittest,
While down the streams of Evolution

> We drift, expecting no solution
> But that of the survival of the fittest.
> Till, in the twilight of the gods,
> When earth and sun are frozen clods,
> When, all its energy degraded,
> Matter to aether shall have faded;
> We, that is, all the work we've done,
> As waves in aether, shall for ever run
> In ever-widening spheres through heavens beyond the sun.[24]

Maxwell's depiction of science as a winged creature shows how highly nineteenth-century scientists valued imagination. Mitchell agreed, and she later wrote that "we especially need imagination in science. It is not all mathematics, not all logic, but it is somewhat beauty and poetry."[25]

In America, Emily Dickinson was one of the best of the nineteenth-century scientific poets. Dickinson used scientific language with force and assurance, probably because she was the only American poet of the time who had received formal scientific training, as a student at the Mount Holyoke Female Seminary in 1847 to 1848. Those were the years when Mary Lyon, president of Mount Holyoke, described the school as a "castle of science."[26] Dickinson's coursework was heavy on mathematics and the natural sciences, and she studied botany and astronomy in great detail.

According to American literature scholar Nina Baym, Dickinson wrote at least 270 poems about science, roughly 15 percent of her total poetic output.[27] Her approach to science was often playful, and Baym argues convincingly that Dickinson's scientific imagination helped bolster her antiauthoritarianism. But Dickinson did not turn to science merely for metaphors; she also wrote about science for its own sake. One poem that would have pleased Maria Mitchell made a careful distinction between astronomy and astrology:

> Nature assigns the Sun—
> That—is Astronomy—
> Nature cannot enact a Friend—
> That—is Astrology.[28]

Precisely because the link between nineteenth-century astronomy and poetry is so close, many of Emily Dickinson's best poems are astronomical. It is unlikely that Maria Mitchell read any of Emily Dickinson's poems, as Dickinson did not start writing prolifically until the 1860s. (There is a small chance that Dickinson's cousins Emily and Louise Norcross showed Mitchell some of their cousin's astronomical poems, as Mitchell was a friend of theirs.) Dickinson probably knew about Mitchell, however, thanks to her own avid interest in astronomy. Astronomers are primed for poetry; both the visuality of astronomical observations and the aesthetic simplicity of astronomical mathematics make it an artistic, even romantic pursuit. The beauty and mythological history of the stars make them perfect subjects for poetry; poems about stargazing have always been common. It stands to reason that as stargazing became more and more scientific in nineteenth-century America, poems about scientific astronomy rather than mere stargazing became more and more prevalent. Drawing on analyses of her poems, as well as on the record of her studies, prominent Dickinson scholars are unanimous in characterizing her as an astronomer. Brad Ricca calls Dickinson a "Learn'd Astronomer," James Guthrie discusses her "long-standing fascination with astronomy," and Richard Sewall notes that "a good deal of astronomy" informs her poetry.[29]

In addition to Dickinson, Edgar Allan Poe and Walt Whitman also turned to science and its language in some of their poems. When Whitman sang the body electric, for example, he was drawing on his knowledge of physiology and physics. Many readers today are familiar with Whitman's "When I Heard the Learn'd Astronomer," a brief poem that expresses the poet's impatience with astronomers' systemization of the stars:

> When I heard the learn'd astronomer,
> When the proofs, the figures,
> were ranged in columns before me,
> When I was shown the charts and the diagrams,
> to add, divide, and measure them,
> When I sitting heard the astronomer,
> where he lectured with much applause in the lecture-room,
> How soon unaccountable I became tired and sick,
> Till rising and gliding out, I wander'd off by myself,
> In the mystical moist night-air, and from time to time,
> Look'd up in perfect silence at the stars.

Edgar Allan Poe's most famous poem about science, "Sonnet—To Science," written in 1829, also objects to systematization. But even Poe's and Whitman's complaints show that poets in the nineteenth century were thinking about science—and many philosophers and poets, Dickinson most emphatically, found science generative rather than vulture-like.

Late in her life, Mitchell confided to a Vassar student that she would rather have written a great poem than discovered a comet. She refused to accept many dichotomies, but the one that outraged her the most was the false opposition set up between scientific truth and aesthetic beauty. She always saw astronomy, and even mathematics, as aesthetically beautiful. She looked for elegance and symmetry in calculations and proofs, for color in the skies. For all the Mitchells, including Maria, astronomy was a science with infinite space for human affection and physical beauty. She certainly objected to Emerson's 1855 lecture at the Atheneum in which he was "very severe on the science of the age. He said that inventors and discoverers helped themselves very much, but they did not help the rest of the world; that a great man was felt to be the centre of the Copernican system; that a botanist dried his plants, but the plants had their revenge and dried the botanist; that a naturalist bottled up reptiles but in return the man was bottled up. There was a pitiful truth in this, but there are glorious exceptions." Training both her poetic and scientific lenses on Emerson, Mitchell went on to write that his speech "was like a beam of light moving in the undulatory waves, meeting with occasional meteors in its path; it was exceedingly captivating."[30]

The importance of the interdependence of science and the humanities in the early nineteenth century cannot be overemphasized; it helps explain why science was less strongly gendered then than it is now. Because the sciences were accessible to all thinkers, not just members of specialized disciplines, there was no sexual binary in the popular mind. After the idea of the two cultures—the sciences and the humanities—became prevalent, each of those two cultures was almost inevitably aligned with a particular gender. For a time, ideas varied on which intellectual realms should admit women; to many educators, the sciences (or at least some of the sciences, particularly biology, botany, and astronomy) seemed feminine. The larger social question of whether women should be educated at all remained.

As we have seen, few Nantucketers debated the importance of women's education. From 1836 to 1857, Maria Mitchell presided over the town's central educational institution, and during this time the Atheneum proudly offered "spheral opportunities" to all. A visitor who climbed the broad stairs and passed through the great white wooden columns of the Atheneum would enter a welcoming place, where "all who choose may gratify their thirst for knowledge."[31] There, surrounded by carefully dusted and arranged volumes, they would see Nantucket's own Athena, Maria Mitchell, who would flash an electric smile of welcome—unless she was deeply involved in reading the latest novel, organizing her ideas about the astronomy of *Paradise Lost,* or calculating a complicated astronomical orbit. Mitchell dedicated herself to the higher things that her Vestal Street home, her challenging schools, her intellectually adventurous church, her scientific and literary clubs, and, most important, her grand and quiet Atheneum encouraged her to pursue.

CHAPTER 3

The Sexes of Science

William Herschel, the British astronomer who discovered Uranus in 1781, wrote so well about astronomy that boys like the young William Mitchell used his books to teach themselves to be astronomers. William Mitchell believed that William Herschel was not only the greatest astronomer of his generation, but also one of the greatest men of all time, and he knew that like himself, William Herschel had never been much of a mathematician. In spite of Herschel's mathematical awkwardness, he had made significant strides toward transforming astronomy from the geographical pursuit of mapping the heavens to the new and far more scientific approach of calculating orbits and gravitational forces.

William Herschel's secret was simple: his sister did the math for him. With an adept and devoted assistant, he attained a level of greatness that would have been completely out of reach for a solitary astronomer. Caroline Herschel worked as William's assistant for forty-one years. Before his appointment as Royal Astronomer in 1781, William had worked as a music conductor, and Caroline had performed with him as a soprano. When William's interests changed, Caroline's apparently changed as well. She taught herself higher mathematics in order to do his astronomical computations for him, and she shared the labor of mapping the heavens, doing all of the calculations necessary to develop their cosmological theories about nebulae and double stars.

Caroline Herschel probably loved astronomy deeply in her own right. She was certainly good at it; she independently discovered three nebulae and eight comets, not the feats of a mere assistant. More important, however, she helped change the course of a discipline: as a result of the Herschels' work, mathematics came to be considered the most important part of astronomy. Observing planets, nebulae, or comets was easy; using math to calculate their irregular orbits and trajectories was, increasingly, the central work of astronomy. Nonetheless, when astronomy textbooks or popular science articles of the day mentioned Caroline Herschel, they focused on her sisterly devotion rather than her scientific passion.[1]

The Herschels' partnership was not an equal one. A devoted servant to her brother, Caroline downplayed her own love for astronomy, writing in her reminiscences: "I am nothing. I have done nothing at all; all I know I owe to my brother. I am only the tool which he has shaped to his use—a well-trained puppy-dog would have done as much."[2] Maria Mitchell shared Caroline Herschel's aptitude for mathematical astronomy, but Caroline Herschel's self-denying attitude was foreign to both Maria and her father. William Mitchell did not think of his daughter as a tool or a puppy, and she would not have accepted that view even if he had. Her father respected her intelligence and wanted and needed her as a mathematical partner on an astronomical team.

Science has always been a highly collaborative, even social endeavor. A glance at the many authors listed on a scientific publication reminds us that scientists don't work alone. This is in part because the Baconian scientific method requires scientists to build on others' results, but also because physical experiments often require many hands. One person starts the ball rolling down the chute, prods the frog's legs, or aims the telescope, another keeps track of the time, and a third records the results. Galileo may have observed alone when he drew the surface of the moon, but he could not have watched the stars, the clock, and his notes all at the same time. Looking from the white-on-black of the telescopic view to the black-on-white of a sheet of recorded observations is hard on the eyes. As far back as 1673, when the Polish couple Johannes and Elizabetha Hevelius used a giant version of a sailor's sextant that was so big it required two people to make an observation, most astronomical observations were made by teams. By the eighteenth century, telescopic astronomy had become a team endeavor, and as astronomy set itself more complex tasks in the eighteenth and early nineteenth centuries, the idea of observing alone became thoroughly obsolete.

At the end of the nineteenth century, Edward Pickering's large-scale photometry program at Harvard, which compared minuscule differences among thousands of photographic plates, employed twenty-three women at the college observatory. In 1921, his successor Harlow Shapley employed forty-two women in a factory-style lab. Shapley explained that his study of variable stars "required a tremendous amount of measuring. I invented the term 'girl-hour' for the time spent by the assistants. Some jobs even took several kilo-girl-hours."[3] The kilo-girl-hour was a unit of measurement that Caroline Herschel, who had put in thousands of hours of her own, would easily have recognized. But the women who worked those thousands of hours were generally not considered astronomers. Shapley blithely remarked, "Luckily, Harvard College was swarming with cheap assistants. That was how we got things done."[4] Harlow Shapley was seen as an astronomer; the forty-two indispensable women who worked in his lab were "cheap assistants." Comments like this one, less than a hundred years old, seem to demand Caroline Herschel–like self-abnegation from women who hoped to do astronomy.

No such demands were made on Maria Mitchell. She was only eleven years old when she began to work as her father's assistant, thirteen when she joined the Nantucket Philosophical Institution, but she was always treated as a valued partner rather than as a cheap assistant. Women who came of age a hundred years later had fewer opportunities and much less encouragement.

It is a strange paradox that the history of science has usually been told in terms of brilliant, solitary men even though the work of science has usually been collaborative. As we reimagine this history, we may find more accurate pictures of science and scientific work. Notably, the cooperation, teamwork, and even self-sacrifice that science requires are qualities that our culture and most others tend to associate with women rather than men. Perhaps our assumptions are wrong on both counts: science is not as "male" as we think it is, and men are not as solitary, unsociable, or self-centered as our culture paints them.

The common stereotype of science as the pursuit of solitary men has its basis in an earlier reality. Early modern scientists believed that science required spiritual isolation: the true scientist needed to deny his body, his passions, and his human circumstances in order to perceive the world without bias.[5] In pursuit of pure observations, early modern scientists struggled to quell the traits they saw as feminine: connections to emotional, physical, and social life. As science became more collaborative, the myth of the scientific mind as isolated and masculine did not diminish; on the contrary, it became more important even as it became more distant from reality.

Eventually the myth reshaped reality. By the mid-twentieth century, women in white coats stood at every laboratory bench in every college and university research lab—the "cheap assistants" were everywhere. But women scientists were regarded as paradoxical beings. Being a scientist meant denying one's feminine qualities to become a disconnected observer; therefore, anyone who became a scientist transcended femininity. By definition, there could be no women scientists.

Twentieth-century attitudes did not spring from nowhere. They harkened back to early modern times, when the study of nature became important to philosophers. But sixteenth-century thinkers were much less literal-minded than Harvard's Harlow Shapley. Although there was a gendered element to the Western cultural formation of the scientific identity in the sixteenth century, as historian Jan Golinski explains, it was philosophically complex: the masculinity of the scientist was defined in terms of self-denial, while the self that each scientist denied was seen as feminine. Associating emotional and domestic ties with femininity is common enough, but a twist in logic is required to frame the scientist's identity as male and define the scientist's self as female.

It may be that early modern thinkers were simply not as rigid about sexual identity as many today. For René Descartes, whose mind-body dualism is central to Golinski's explanation of the masculinization of science, all bodies were feminine, whether they were male or female. Reason—the mind—was masculine, regardless of the sex of the body that housed it. Descartes thought that generally men tended to be more rational than women, who were often more physical, but he respected individual women's intellects. Many of his best arguments were formed in conversation with Princess Elizabeth of Bohemia, one of his closest friends and most respected interlocutors.[6] Elizabeth questioned Descartes' denial of the body right at the start, and Descartes was hard pressed to counter her arguments; they corresponded about the mind-body split in the early 1640s. But then, ironically, Descartes went to Sweden to tutor the rigidly self-denying Queen Christina, and her ascetic physical regime finished him off. Forced to rise at five a.m., take cold baths, and eat scanty meals, the philosopher perished from the hardship. Queen Christina was much better at denying the body than Descartes, and her body was better able to withstand such treatment than his. Descartes would probably have been startled to learn that his masculine and feminine metaphors would one day be misunderstood as physiological realities. Although he equated the scientist's and the philosopher's self-denial with masculinity, that doesn't mean he thought that only men could master it.

Princess Elizabeth and Queen Christina are wonderful examples of women in early modern natural philosophy, as are two British noblewomen devoted to scientific inquiry, Ann Conway and Margaret Cavendish, who were able to use their money and social position to build communities of researchers with themselves at the center. The reason all these stories are remembered is that they concern extremely wealthy noblewomen: countesses, princesses, and queens. Those archival records that survive from the sixteenth to eighteenth centuries rarely mention peasants or tradespeople of either sex, so it is hard to determine the attitudes of more common women and men toward scientific inquiry. In England and Europe, early modern science was even more of an aristocratic endeavor than it was a male one. This is an important point, because it explains why science was a thoroughly amateur activity until the late eighteenth century: wealthy aristocrats pursued scientific knowledge for its own sake, with no thought of trying to earn money from their work. Because science was not a profession, it had more social prestige than law or the church. Those who needed to support themselves might train as clerics or lawyers or even soldiers or sailors; those with independent means were free to turn to private vocations such as natural philosophy.

Still, it was generally not a solitary hobby. Teamwork was essential to operate experimental contraptions, so countless servants and family members were drawn into the practice of science. The names—and the ideas—of most of the servants are lost to history, but family assistants have left helpful traces of their activities. In the case of astronomy, the most prominent family teams before the Herschels and the Mitchells were probably the husband-and-wife teams of Johannes and Elizabetha Hevelius, in Poland, and John and Margaret Flamsteed, in Britain, but the annals of early chemistry, botany, geology, and zoology are replete with devoted sisters, daughters, and wives. As the historian Sally Kohlstedt explains, the importance of family connections in early American science "is demonstrated in 'scientific genealogies' whose branches reach upward across generations and outward through kinship networks. Best recorded are immediate linkages between parent, specially father, and child. ...."[7]

Closer to Nantucket, in Cambridge, Massachusetts, the Harvard astronomer William Bond worked so closely with his son George that they built their great refracting telescope with a two-person seat attached to the telescope mount; this way, they could always observe together, sitting in comfort on a Victorian love seat upholstered in fine red wool. When Maria Mitchell and her father visited the Bonds, two of the rare souls who shared their love for and expertise in astronomy, Maria shared the telescope couch with her father or her friends.

Caroline Herschel and Elizabetha Hevelius were as much astronomers as William Bond's lifelong assistant, his son George, was. They stayed up late into the night observing the skies; they took notes, kept detailed records, and often did the math. Some family assistants might have been motivated by love for a man rather than love for a subject, but regardless of their motivations, they made important contributions to science. In addition to family members, there were enough women who came to astronomy independently that the muse of astronomy, Urania, was often imagined as a woman astronomer. When the Swiss astronomer Marie Cunitz published *Urania Propitiae,* a useful revision of Kepler, in 1650, the title page of her book was designed so that her name was equated with the muse's.[8] By 1786, there had been enough women astronomers for Jérôme de Lalande, chair of astronomy at the Collège de France and an enthusiastic supporter of women in astronomy, to publish a book chronicling their achievements. De Lalande had calculated the orbit of Halley's comet with Nicole-Reine Lepaute, and when his daughter was born, he named her Caroline in honor of Caroline Herschel.[9]

This is not to say that women have had the same scientific opportunities as men in any era; they haven't. Nor does it suggest that the work women and servants did in the seventeenth and eighteenth centuries was the same as that performed by the aristocratic men who led most scientific inquiries; it wasn't. But some of those women were far more than assistants, and I personally would call them scientists.

Although scientific workers who were female were rarely thought of as scientists in the seventeenth and eighteenth centuries, those who were American were even less likely to be considered scientists. The situation in the United States was quite unlike Europe's, not least because the country did not have an aristocracy parallel to Europe's. Throughout the eighteenth century, scientific inquiry in America was associated with the leisure activities of clergymen, such as Jonathan Edwards, and politicians, such as Benjamin Franklin and Thomas Jefferson. American approaches to science were more likely to take a pragmatic turn: Jonathan Edwards was interested in lightning, fire, and optics, all of which he turned toward theology and used in his sermons and meditations. Franklin's interest in electricity begat the lightning rod, while his experiments with fire led to the Franklin stove, and optics led him straight to bifocal lenses. Jefferson's *Notes on the State of Virginia* was more compelling for its political purpose than for its scientific observations. Franklin is the best example of an early American scientist, but he had hundreds of other, more pressing obligations. Science was a hobby for him, an avocation that might help humanity by proposing practical devices. He was as committed to his amateurism as any British lord.

Perhaps because Franklin did not see science as a profession, he had no doubt that it was an appropriate subject for girls. But when Franklin wrote letters to his dear Polly Stevenson about "natural philosophy," he did not mean exactly what we mean by "science." Natural philosophy required rigorous thinking about natural phenomena, but most natural philosophers worked independently, doing experiments they funded themselves and sharing their results with like-minded amateurs. They did not work in institutions. The few who were wealthy and leisured enough to define themselves in terms of these interests were most likely to style themselves "philosophers" and to think of their interests as high flown rather than practical. Natural philosophy in the seventeenth and eighteenth centuries was generally unpragmatic and nonmercantile—closer to metaphysics than to mechanics. Science as we know it did not really exist before 1800. The historian David Cahan reminds us that it is only in "the final third of the nineteenth century" that we can "speak legitimately, that is, in a modern sense, of 'science,' 'scientists' and the discipline of science."[10] Before then, "science" was just another word for specific knowledge, not a description of a discipline, and "scientists" did not to define themselves or their work as part of a coherent discipline that was somehow opposed to the humanities.

In the late nineteenth century, many paradigms shifted, not only for scientists but also for the definitions of "man" and "woman." Maleness and femaleness have had different social, scientific, and even physiological meanings at different points in history, and now Anglo-American medicine and culture began to move from an understanding of sexuality as something amorphous and undifferentiated toward an almost fanatical belief in "true sex" and in sexuality as an identity.[11] For Descartes in the seventeenth century, every person was both male and female, mind and body. This attitude, which the scholar Thomas Laqueur calls the one-sex model, was still prevalent in Benjamin Franklin's day.[12] Most people shared the understanding of physiology expressed in *Aristotle's Masterpiece,* the popular sex guide and physiology book.

> That, tho' they of different Sexes be,
> Yet in the whole they are the same as we:
> For those that have the strictest searchers been,
> Find Women are but Men turned outside in;
> And Men if they but cast their eyes about,
> May find they're Women, with their inside out.[13]

Because of the morphological similarity between ovaries and testicles (and the less convincing similarity of the penis to the uterus), Europeans and European Americans had long believed that women and men were physiologically the same, and even that it was possible to slip from one to the other. The most ridiculous supposition was that if a girl jumped over a high fence at the wrong angle, her genitals could fall out, thus turning her into a boy, her inside-out uterus becoming a penis. A more likely prospect was that a woman might play the social role of a man; for example, when her husband was away on a long sea voyage. Although early Americans had very definite ideas about sex roles, they were sometimes surprisingly open-minded about sex identity—if circumstances required it, a woman could become a man in her social role, if not anatomically. But that tolerance would change in the nineteenth century, when the social aspects of sex identity became fixed.

At the close of the eighteenth century, there was a move away from identifying the intellect as male. During the Age of Reason, many people began to believe that women were reasonable beings, and indeed might have intellects very similar to men's. In her controversial treatise *A Vindication of the Rights of Women,* Mary Wollstonecraft expressed the increasingly acceptable notion that "the mind has no sex." That was perhaps the least controversial of her opinions, for Wollstonecraft was a revolutionary. In 1788, when she was about the same age as Walter Folger, Wollstonecraft decided to leave her position as a schoolmistress in England and travel to Paris. Along the road, hers was the only carriage moving *toward* the city; everyone else who could afford to hire a carriage was in flight from it.

In revolutionary Paris, Wollstonecraft found the freedom and ideals she had longed for. She wrote indefatigably, and the London papers were hungry for her news, as few English speakers had had the courage to stay in Paris. While she was there, Wollstonecraft fell in love with an American and had a child with him. She did not marry but lived by herself in a cottage and supported herself with her ideas and her observations. The market for her words was vast. When Wollstonecraft returned to England, she wrote *A Vindication of the Rights of Women,* in which she argued passionately for women's rights to education and entry into the realm of ideas and politics.

Wollstonecraft was resolutely single. Even after she fell in love a second time, with William Godwin, she did not give up her lodgings or her writing. She and Godwin traded manuscripts and printers' proofs, made love in her apartment, ate in tea shops and pubs,

and endlessly bantered and talked. When Wollstonecraft was preg-
nant with her second child, they finally married. On the day her
daughter was born, one of Caroline Herschel's comets flamed over-
head. The child was named Mary Wollstonecraft Godwin, and she
would later (as Mary Shelley) become a distinguished writer in her
own right, offering an extended discussion of mid-nineteenth-
century science in her most important novel, *Frankenstein*. But Mary
Wollstonecraft, tragically, did not survive childbirth. The *Vindication*
was rushed to print, and Godwin forsook his own work for a time to
produce and publicize it.

Meanwhile, in a drafty manor in Ireland, Maria Edgeworth was
fighting a quieter fight for women's educational rights. Edgeworth's
father, William, was one of the so-called Lunar Men, a small club of
merchants and thinkers devoted to natural philosophy. Maria performed scientific experiments with her father and wrote her first
books as his assistant in addition to raising her eleven young siblings.
She taught all of them to think: to read and calculate, to classify
plants and birds, to consider the physics of light gleaming through
the cut facets of the crystal decanter, to wonder about the chemistry
of rising bread in the kitchen. Edgeworth's books were much quieter
than those of Wollstonecraft. Rather than insisting that women have
minds and must be educated, Edgeworth simply assumed it and
began the practical work of education. Presuming, correctly, that
most of her readers were women, she aimed her books at a double
audience: women educating girls and boys, and the girls and boys
themselves.

Both Mary Wollstonecraft and Maria Edgeworth loved the world
of ideas, and both believed that women were thinkers who had the
right—indeed, the responsibility—to educate themselves. But they
had very different personalities. Wollstonecraft was daring, tempes-
tuous, and confrontational; she made a scene wherever she went.
Edgeworth was far more domestic; she seems to have spent more
time thinking about her young siblings than dreaming of possible
lovers. She never married but spent her life caring for her younger
siblings. Even more crucially, Edgeworth was not a revolutionary
where political rights were concerned. She concerned herself with
education, not politics, and, unlike Wollstonecraft, she seems to have
believed that the two could remain separate.

In the United States, Edgeworth's model took hold. Bookish

families such as the Mitchells learned her stories by heart. By comparison, Wollstonecraft terrified many Americans. Her public, unmarried sexuality and undomesticated living arrangements shocked many sensibilities. With a flexible sense of sexual identity left over from an earlier era, the poet and clergyman Richard Polwhele went so far as to call her an "unsex'd female." Still, in spite of the controversies that swirled around Mary Wollstonecraft's personal life, her idea that the mind had no sex—that women possessed intellects similar to men's—took hold in America.

Education in America changed drastically during this time. In 1800, few Americans were formally educated, and even fewer had college or university degrees. By 1900, colleges and universities had replaced high schools as the sites of higher education, and an educated person was one who had attended a college or university. Although this change was positive for women in many ways, it turned out to be a mixed bag for scientific women. Once colleges came to dominate the educational scene, there were fewer opportunities for women to study and teach science.

The path of the woman scientist must be traced through all of these shifting paradigms. The intertwined histories of science, sex, and education in the early nineteenth century show that pre-institutional science was one of the intellectual realms that welcomed women most enthusiastically. But this early history also foreshadows the sad change of tides that would push women out of science almost completely.

CHAPTER 4

Miss Mitchell's Comet

After the Mitchells moved to the Pacific Bank at the top of Main Street in 1836, things were much less crowded. Maria had her own room, and she had the vast Grecian expanse of the Atheneum to herself all morning. The bank's rooftop observatory was not as small or angular as the old one on Vestal Street had been—the roof of the bank was flat—so Mitchell climbed "up-scuttle" to her rooftop observatory every clear night for more than fifteen years. If it was cold, she wrapped herself in a heavy wool coat. Sometimes snow drifted around the instruments; sometimes spiders disturbed her solitude. She always carried a small observing notebook and a little whale-oil lamp that illuminated only her notes. From 1836 to 1847, her routine changed very little. Mitchell confessed in her journal that she felt attached "to certain midnight apparitions. The aurora is always a pleasant companion, a meteor seems to come like a messenger from departed spirits and even the blossoming of the trees in the moonlight becomes a sight looked for with pleasure. And from astronomy there is the enjoyment as a night upon the housetops with the stars as in the midst of other grand scenery. There is the same subdued quiet and grateful sensuousness—a calm to the troubled spirit and a hope to the desponding."[1]

On October 1, 1847, at the age of twenty-nine, Maria Mitchell excused herself from a dinner party, wrapped herself up in her coat, and climbed out onto the roof of the Pacific Bank. She had been slowly and methodically "sweeping" certain quadrants of the sky with precise regularity for years, searching for something new. That night she saw it: a blurry streak of moving light that could only be a comet. She called her father from the sociable group below, and both of them were delighted.

It might seem easy to an untrained onlooker, but discovering a telescopic comet requires long-term diligence. In order to recognize a tiny blur of light as a new comet, the astronomer needs to know precisely what was there before. Sweeping the skies takes painstaking precision, and discoveries take a dose of pure luck. George Bond had swept the same region of the sky just a few weeks earlier without finding anything new; in Europe, Frau Rümker in Germany and Father Secchi in Rome each discovered the comet independently within days of Mitchell. A fortuitous mix of talent and circumstance pushed her into the limelight.

Centuries before 1847, a small comet a few kilometers in diameter hurtled through the universe, making rapid progress on a nearly straight course toward our sun in the Milky Way. The winter of 1847, after Maria Mitchell had spotted the comet with her telescope and alerted the world to its approach, astronomers watched Comet Mitchell as it made a sharp turn around the sun and sped off again into the infinite reaches of the universe. The comet's path through space was irrevocably altered by its curve around the sun that year— and so was Maria Mitchell's.

Mitchell had already established herself as a serious astronomer on the national level. When Benjamin Pierce, the Harvard professor of mathematics, had recommended her father for a job, he had singled out Maria as "an accomplished mathematician and excellent observer."[2] Mitchell's friend and early biographer Julia Ward Howe commented, "As years passed on, Miss Mitchell began to be spoken of as a woman of uncommon merit and attainments. [I] remember to have heard of her as an astronomer of recognized position as early as the year 1846. She was living at Nantucket at this time, and had probably no anticipation of the publicity about to be given to her modest and quiet labors."[3] But astronomy in America was in its infancy, and no American astronomer had much reputation in Europe. A few American astronomers, including William Bond, had discovered comets, but none had been able to establish the priority of their claims.

Ironically, prestige had found an unwilling subject. All of the Mitchells were hesitant about making public spectacles of themselves, in part because of their Quaker beliefs. Maria's sister Anne stressed the family's shared modesty when she recounted the discovery:

> As soon as tea was over she said to them: "Now, you must excuse me. The heavens are so clear I want to sweep the skies. Who knows what comets may be roaming at large?" After about an hour we heard my father running quickly down stairs. He opened the parlor door, his observing cap down to his eyes, and exclaimed, "Maria has found a telescopic comet!" The general rejoicing of the guests contrasted oddly with the quiet demeanor of the mother and sisters. When Maria heard the stir of departing guests and came down to say good-night, her friends clustered around her with congratulations. "It was there," she said simply; "How could I help seeing it? There was no merit in that."[4]

Maria objected when her father, William, wanted to publish her claim to discovery. William wrote to Edward Everett, the president of Harvard,

> I urged very strongly that it be published immediately, but she resisted very strongly, though she could not but acknowledge her conviction that it was a comet. She remarked to me, "If it is a new comet, the Bonds have seen it. It may be an old one, so far as relates to the discovery, and one which we have not followed." She consented, however, that I should write to William C. Bond, which I did by the first mail that left the island after the discovery.... Referring to my journal, I find these words: "Maria will not consent to have me announce it as an original discovery."[5]

Yet in spite of Maria's reluctance to lay claim to the discovery, the Bonds and all of her astronomical acquaintances were delighted to learn of it. They were convinced, even if she was not, that she was an original discoverer. She treasured the joking letter from her friend George Bond, his father's assistant at Harvard: "If you are going to find any more comets, can you not wait at least until they are announced by the proper authorities? At least, don't kidnap another such as this last was. If my object were to make you fear and tremble, I should tell you that on the evening of the 30th I was sweeping within a few degrees of your prize . . ."[6]

The letter from Alexander Dalles Bache, the director of the United States Coast Survey, addressed to "the lady astronomer in whose fame I take personal pride," was equally merry: "We congratulate the indefatigable comet seeker most heartily on her success; is she not the first lady who has ever discovered a comet? The Coast Service is proud of her connection with it! Now if she determines the orbit also, it will be another jewel for the civic crown. As far as to feathers and caps, you eschew all others."[7] In his reference to "feathers and caps," Bache teased Mitchell for her Quaker-style modesty even as he urged her to claim the discovery for the sake of "the civic crown." His strategy of appealing to her on behalf of American astronomy showed great insight into her character.

By November 12, 1847, Maria had published a preliminary notice of her discovery in the *Monthly Notices for the Royal Astronomical Society* in Britain. In January 1848, an article on "the comet of the tenth month," as she and her father discreetly called it, was published under William Mitchell's name in *Silliman's Journal* (later the *American Journal of Science and Arts*). Finally, on February 11, 1848, the *Monthly Notices for the Royal Astronomical Society* contained an article on "Miss Mitchell's Comet," as everyone but the Mitchells insisted on calling it. In that article, she followed up on Bache's advice and published a calculation of her comet's orbit. By doing the math, she made the discovery hers. When she published her results in *Silliman's* and the *Monthly Notices,* she consented to the public recognition from which she had originally recoiled.

Within weeks, the entire American astronomical community joined in the Mitchells' excitement. Prompted by the Bonds, Edward Everett, president of Harvard University, went so far as to publish a notice about her comet in *Astronomische Nachrichten,* an international journal published in Germany, as part of his campaign to convince the Danish government to award Mitchell their gold medal for astronomical discovery.[8] Everett also wrote to the Danish consul in Washington to argue Mitchell's case: "As the claimant is a young lady of great diffidence, the place is a retired island, remote from all the high-roads of communication; as the conditions have not been well understood in this country; and especially as there was substantial compliance with them—I hope his Majesty may think Miss Maria Mitchell entitled to the medal."⁹ The king of Denmark, who was the final arbiter because he had established the prize, was duly convinced that Mitchell had discovered the comet days before anyone in Europe had. The Danish government awarded her the gold medal, making her the first American astronomer to be formally acknowledged by the European scientific community for an original discovery. She had now bolstered the case for all American astronomers, who longed to be on an equal footing with the observatories of Denmark and Britain. She also captured the imagination of a nation increasingly enthralled with discovery and celebrity. Mitchell was no longer merely an astronomer; she was a star.

The discovery of a comet was momentous for Mitchell and her reputation, but it is important to understand what her discovery meant. Throughout human history, comets had been linked to political revolutions. For thousands of years, revolution was seen as a negative force, but by the 1840s the concept had acquired some positive associations. Symbolically, to many eyes, a "woman's comet" signified new political possibilities for women. Comets also had some political meaning for astronomy as a discipline. The great comet of 1843, whose brilliant tail had filled the sky with light, was so spectacularly popular that it convinced the American government to invest in astronomical research. At the same time, comets had purely astronomical significance: astronomers pursued comets because of what they could teach humanity about irregular orbits.

Maria Mitchell's comet was significant in all of these ways. Her discovery was trumpeted at the 1848 Seneca Falls Convention for women's rights, but the public was at least as excited by the fact that an American had discovered a comet as by the fact that a woman had done so. Astronomically, Mitchell's comet was notable because it was a hyperbolic comet whose orbit was a challenge to calculate. When she published her calculations in the *Monthly Notices of the British Royal Astronomical Society,* her achievement impressed three crowds at once: those interested in women's rights, in American science, and in pure astronomy.

From classical times up through the mid-seventeenth century, comets were regarded as portents of disaster. In fact, the word "disaster" comes from Latin roots meaning "evil star." The good stars were the regular, well-behaved ones that revolved around the axis of the pole star, securely fixed in their own constellations. The rest were bad stars: meteors, asteroids, and comets that showed up at random and flew unpredictably through the heavens. Comets were the worst kind of star—not only did they move in wildly unexpected ways, but also they sometimes hung around for months at a time, upsetting the movement of the celestial spheres, not to mention their human observers. In eras that respected the natural order of things, comets were evil omens of the disturbance of the natural order—disasters. With their tails blazing across the sky, they seemed much larger and more significant than the other stars. In particular, they were seen as portents of the death or dethronement of kings.

In classical times, comets were associated with the violent death of rulers. Two thousand years ago, the Greek historian Plutarch related that a comet had appeared in the sky when Julius Caesar was assassinated, in 44 BCE. "The most signal preternatural appearances were the great comet, which shone very bright for seven nights after Caesar's death, and then disappeared."[10] Comets continued to symbolize violent political changes for more than a thousand years; in the eleventh century, the women who embroidered the Bayeux tapestry commemorated the great comet of 1066, later known as Halley's comet. The comet had lit up the sky for months during that eventful year when the English nobleman Harold deposed Edward the Confessor, defeated the Viking Harald, and then was himself deposed by William the Conqueror. But the comet's meaning was open to interpretation. The tapestry artists chose to link the comet to Harold, who had deposed one king and then been deposed himself; they embroidered the comet above his head. Like a comet, Harold had rebelled against divine order: he had tried to disrupt the royal succession.

Another example of the link between comets and dethronements comes from Shakespeare. In the late sixteenth century, Shakespeare relied on Plutarch's description of the comet associated with Julius Caesar's assassination. In *Julius Caesar,* Caesar's wife, Calpurnia, explains:

> When beggars die there are no comets seen
> The heavens themselves blaze forth the death of princes.[11]

Calpurnia's comment shows that the association of comets and royal troubles, which had seemed transparently obvious from the age of Caesar up to the Norman conquest, had grown a bit more complicated by Shakespeare's time. In the early modern era, some people began to question whether the actions of kings were, in fact, as divinely ordained as those of the stars. When John Milton wrote of comets in *Paradise Lost,* he gave them two different shades of meaning. First, he linked comets to Satan, who had attempted to dethrone God and was about to take on Death itself. When Satan and Death face each other at the gates of Hell, Milton compares the "unterrified" Satan to a comet that "from his horrid hair / Shakes pestilence and war."[12] But although Milton loosely upheld the association between comets and rebellion against authority (and the calamities, such as pestilence and war, that go along with revolutions), his attitude toward the dethronement of kings was not necessarily negative. During the English civil war, Milton wrote a justification of rebellion against kings and accepted a cabinet position under Oliver Cromwell, the Puritan leader of the opposition to King Charles. The second and final comet image in *Paradise Lost* occurs at the very end of the poem, when Adam and Eve are banished from Paradise: God's flaming sword is described as "fierce as a comet" and is clearly intended as a symbol of justice, not evil.

Milton's tweak of the comet's symbolism may be related to his own revolutionary politics, or to his fascination with Galileo's astronomy. (Maria Mitchell wrote a long essay on Milton's astronomy, discussed elsewhere.) But Milton was also a creature of his time, which looked more favorably on revolution than previous eras had. One hundred years later, by the late eighteenth century, many Enlightenment thinkers were advocates for revolution, and revolutions in America and France had succeeded in removing or disempowering kings. At this point, comets had become much more ambiguous symbols: they were still widely associated with rebellion, but rebellion itself had become a social positive.

In 1786, when the English writer Fanny Burney delighted in her view of "the first ladies' comet" (discovered by Caroline Herschel), she identified comets with women's rebellion against male dominance. As one of the queen's ladies-in-waiting, she was there when William Herschel came to the palace to set up his telescope so that the royal women could view the comet his sister had discovered. Ten years later, Mary Wollstonecraft gave birth to her daughter Mary (who would become the writer Mary Shelley) while another of Caroline Herschel's comets blazed in the sky. Daughter Mary was convinced that the comet was a good omen and that she had been born under a lucky star. In her first collection of poems, she apostrophized:

> And thou strange star! Ascendant at my birth,
> Which rained, they said, kind influence on the earth.
> So from great parents sprung, I dared to boast
> Fortune my friend.[13]

When Mary Shelley, the daughter of radical thinkers, associated comets with "kind influence on the earth" rather than with Milton's "pestilence and war," she did so because she approved of social change. Later, in *Frankenstein,* Shelley would reveal a more ambivalent attitude toward modernity, and particularly toward the developing sciences. But as a girl in the opening years of the nineteenth century, she was taught by her feminist father that comets were lucky stars, predicting women's new equality.

The first "ladies' comet" was one of many. The German astronomer Maria Kirch was probably the first woman to discover a comet, in 1702. Nicole-Reine Lepaute did not discover any comets, but she did the laborious calculations of the orbit for Halley's comet in 1759. In Italy in 1854, Caterina Scarpellini discovered a comet. And even if Maria Mitchell had stayed inside on that chilly October evening in 1847, her comet was destined to be a ladies' comet, as the next person to see it was Frau Rümker, wife of the director of the Hamburg observatory in Germany.

Comets were central to nineteenth-century astronomy. When Caroline Herschel discovered her first, Jérôme Lalande wrote to congratulate her, saying, "At the moment, comets are what interest us most. We are expecting several from you."[14] Lalande's expectation was not just a sign of his feminist sympathies; searching for comets was exactly the sort of astronomical work that he felt women were suited for.

Discovering telescopic comets did not require complicated equipment. On the contrary, a large, cumbersome telescope was something of a disadvantage in the process of sweeping the skies; a small, maneuverable instrument was ideal. Even today, although automated computers discover most comets, those discovered by humans are very often found by amateurs. The two men who discovered the Hale-Bopp comet of 1997 were both using their own telescopes and working outside institutions—Hale was an unemployed astronomer with a PhD, and Bopp had no formal training. The most successful comet-finder in history is Carolyn Shoemaker, born in 1929, an astronomer with no formal training who has discovered thirty-two comets to date, including Shoemaker-Levy 9, the spectacular comet that exploded into icy fragments when it nearly collided with Jupiter in 1992.

In 1848, the British royal astronomer John Herschel tried to explain why comets were so interesting to astronomers: "The extraordinary aspect of comets, their rapid and seemingly irregular motions, the unexpected manner in which they often burst upon us, and the imposing magnitude which they occasionally assume, have in all ages rendered them objects of astonishment, not unmixed with superstitious dread to the uninitiated, and an enigma to those most conversant with the wonders of creation and the operations of natural causes."[15] Herschel went on to explain, "In fact, there is no branch of astronomy more replete with interest, and, we may add, more eagerly pursued at present."[16]

Then as now, comets were extremely mysterious. No one was quite sure what a comet was, or whether it generated its own light or merely reflected stellar light. No one knew what started a comet on its erratic orbit. Some comets are tied to our solar system by elliptical orbits around the sun, but many others merely pass through on parabolic or hyperbolic paths. (The hyperbola and the parabola are both variations of an open elliptical path—not unlike a single parenthesis—while a comet that stays in our solar system has a closed elliptical orbit around the off-center focal point of our sun.) A parabola is more sharply curved than a hyperbola; its arms seem to be forming an oval, while the hyperbola's arms are so minimally curved that it is almost a straight line. These nonelliptical comets never return; they travel from deep space toward the sun, curve around it, and then spin out again into the far reaches of the universe. No one knows exactly what brings them into our solar system.

But even though comets remain among the most enigmatic objects in space, they have also helped to solve some significant astronomical problems. Of all the comets in human history, Halley's is probably the most scientifically significant. Sir Edmund Halley, a friend of Sir Isaac Newton, figured out that the comet he saw in the English skies in 1680 had been there before. He observed the comet long enough to calculate the geometry of its orbit and decided that it was an elliptical comet on a seventy-eight-year orbit around the sun. Combing the historical record, he found evidence of significant comet sightings every seventy-eight years, including the comet that had blazed above England and France during the turmoil of 1066. He predicted that the comet would return in 1758. When it did, modern astronomy was born.

The return of Halley's comet marked a significant turning point in astronomy for several reasons. Halley's geometrical calculation of the comet's elliptical orbit showed that geometry could work as a predictive and investigative astronomical tool, not just as a descriptive tool. Newton had described known solar orbits using mathematics, but Halley went one step further and used Newton's mathematics to predict an unknown, as yet unobserved orbit. In 1852, Robert Grant's *History of Physical Astronomy* remarked that "the fidelity with which [Halley's comet] responded to the deductions of the geometer on the occasion of its last appearance forms one of the many magnificent triumphs which adorns the Theory of Gravitation."[17] When the comet returned in the middle of the eighteenth century, astronomers were able to elaborate on Halley's calculations to account for the gravitational effects of the large planets the comet passed. The French team of Clairault, Lalande, and Lepaute successfully calculated the orbit's complexities, and Nicole-Reine Lepaute was said to have borne the brunt of the laborious calculations—making her one of the great female contributors to comet astronomy. Halley's comet helped to change astronomy definitively from an observational science to a mathematical one, in which geometry was used not just to describe observations but to explore and predict future events. From 1759 onward, astronomy was more akin to astrophysics than it was to stargazing.

Comet astronomy also changed astronomers' understanding of the vast distances of space, and thus of the function of telescopes. When Galileo began to use a telescope for astronomical observation in 1609, no one was sure how to explain the instrument's optics. Everyone soon agreed that the telescope made faraway objects appear closer, but it was not known whether telescopes merely brought blurry distant objects into sharp focus or actually extended the range of vision. Premodern astronomers had no way of judging the distance of an object: a meteor blazing quickly across the sky as it burned up in the earth's atmosphere was hard to distinguish from a comet thousands of miles away. William Herschel believed that telescopes actually allowed observers to "penetrate into space"—to see at a greater distance. The known planets were of no use in proving this because they were always in sight, and the stars were equally unhelpful because they stayed at roughly the same distance from the earth. But comets traveled into and out of the range of sight. Comets that could not be seen without a telescope were known as telescopic comets. As they traveled into the range of human vision, telescopic comets taught valuable lessons on how a telescope worked and offered fascinating insights into the dimensions of space. Maria Mitchell's hyperbolic comet of 1847, for example, will never come back into our solar system, but because telescopes get better and better, that comet has not left our telescopic sights. In 1997, 150 years after Mitchell first saw the comet, astronomers affiliated with the Maria Mitchell Observatory on Nantucket reported to the American Astronomical Society that telescopes could still find the comet, commenting, "The comet is 32(m) [a very small magnitude of brightness] now and will slowly decrease in brightness. Our ability to detect faint objects has improved dramatically over the 150 years since the comet was discovered. By extrapolation, we show that modern technology may catch up with the declining brightness of the comet by the middle of the next century. Another challenge for astronomers!"[18]

It makes sense for professional astronomers to leave comet discoveries to the amateurs or, in recent decades, to computers. It is rote work, and it is rarely successful. Caroline Herschel discovered eight comets, but she swept the skies nightly for forty years. Maria Mitchell swept for comets even longer and made only one original discovery—although she independently found eight other comets before anyone else had announced them, 1847 was the only time she qualified as a "first discoverer," the one who had seen the comet before anyone else on earth. She managed to find the other eight before their discoveries had been announced, but she was not the one who saw them first, that is, the first discoverer—a distinction of great importance to astronomers. Maria's brother Henry commented that "sweeping the heavens with the telescope through the long hours every clear night, as Maria Mitchell was wont to do, means healthy courage and hopes that prophesy success."[19] The patience this work requires cannot be overestimated, but the relatively simple equipment makes it an ideal venture for people on the edges of institutional science: women, amateurs, and, in the nineteenth century, Americans.

Since 1945, the United States has been at the forefront of scientific research and discovery and a world leader in astrophysics. But one hundred years earlier, in the 1840s, the United States was far behind Europe. Going back two hundred years, to the 1740s, the British colonies were about on par with England and other European countries, but in the mid-eighteenth century, the Europeans began to institutionalize their scientific practice, beginning with England's Royal Society and the Institut de France. In other European courts, heads of state placed scientific inquiry, and particularly astronomical observation, under governmental oversight. There were state-sponsored observatories in Denmark, Germany, and Russia, as well as in England and Scotland. For about a century, the United Stated lagged behind.

When Maria Mitchell discovered the comet, she, like most American astronomers, was an amateur. William, who drew a salary of one hundred dollars a year from the United States Coast Survey, was among the few who were paid anything for astronomical labors; even William Bond, director of the Harvard College Observatory, saw his astronomical work as more of a hobby than a profession. From 1839 to 1846 he did not draw a salary from the college but supported himself by designing and selling chronographs and other instruments, some of which he used in his observations. His association with Harvard was fairly tenuous; occasionally he'd let the college boys peer through one of his telescopes. Gradually the college took more of a role, and in the mid-1840s, when Harvard donors funded the purchase of a great refractor that rivaled those in Europe, the relationship between the observatory and the college became more formal. In 1846 Bond was finally granted a salary, and he was paid until his death, in 1860. William Bond was never appointed to a professorship at the college. His son George Bond earned a Harvard degree and worked as his father's unpaid assistant for decades before he was promoted to Phillips Professor of Astronomy in 1860.[20]

Like the observatories in Cincinnati and Washington, D.C., the Harvard Observatory was spurred on to great improvements by the appearance of the great comet of 1843, which made Americans aware of their astronomical deficiencies. The comet reached at least the third magnitude of brightness; one observer from a whale ship reported that when he watched the comet pass in front of the sun, it was visible the whole time. Its tail was more than two hundred million miles long—about the distance from the sun to Mars—and one shipboard observer reported that the comet reached more than a third of the way across the night sky, filling 70 degrees of the possible 180 degrees.[21] And yet, in spite of the comet's massive and brilliant grandeur, the Americans failed to record any substantive observations of it. They simply didn't have the instruments at the time, and there was no official national observatory. The best data on the comet came from student observers at Philadelphia High School. Maria Mitchell was so frustrated by the inadequacy of her and her father's observations of this comet that she removed them from their otherwise complete records, leaving a note that said: "In this portion of this book I found a great many scraps about the comet of 1843. I have removed them. Maria Mitchell."[22] The Bonds at Harvard were equally frustrated.

This failure motivated President John Quincy Adams, a longtime advocate for a national observatory, to deliver an oration on the importance of astronomy, saying in conclusion:

> When our fathers abjured the names of Britons and "assumed among the powers of the earth, the separate and equal station, to which the laws of Nature and of Nature's God entitles them," they tacitly contracted for themselves, and above all for their posterity, to contribute...their full share...of the virtues that elevate, and of the graces that adorn the character of civilized man.... But have not the labors of our hands, and the aspirations of our hearts been so absorbed in toils upon this terraqueous globe as to overlook its indissoluble connection...with the firmament above?[23]

But Congress resisted Adams's initiatives, and clever advocates manipulated them. In the early 1840s, lobbyists for the U.S. Navy succeeded in gaining congressional support for funding a small observatory that would be attached to the navy's Depot of Charts and Instruments. George Bancroft, the famed historian and secretary of the navy, agreed with Adams that a national observatory was important; building on the fact that the depot already existed and was stocked with astronomical instruments that were useful for navigation and for oceanography, he skillfully massaged the anti-Adams factions of Congress into funding a naval observatory that would supersede Adams's vision for a national facility. By 1846, Adams was forced to concede. "I am delighted," he wrote, "that an astronomical observatory—not perhaps as great as it should have been—has been smuggled into the number of institutions of the country under the mask of a small depot for charts."[24]

The small observatory at the U.S. Navy Depot was in no position to compete with national European observatories. The Harvard Observatory, despite having the best telescope in the United States, which cost twenty thousand dollars (perhaps the equivalent of a few million dollars today) and was on a par with the world's largest up to that point, was still not recognized as a formal national institution. It took Maria Mitchell's discovery to bring American astronomy onto the world stage.

Across the board, American science was attempting to professionalize. When Alexander Dalles Bache of the U.S. Coast Survey argued that the United States should establish its own nautical almanac, his primary reason was that American astronomy needed to be official in order to keep up with government-sponsored European astronomy. The U.S. Coast Survey was mainly concerned with oceanographic data: charting currents and coastlines, ocean depths and wind patterns. Astronomy was an important element of the survey because the charts required exact longitudes and latitudes. William Mitchell and his assistant Maria worked for the coast survey through the 1830s and 1840s, taking the careful readings that helped to establish longitudinal meridians. American astronomers longed to set astronomy on a firm footing in the United States, but this was surprisingly difficult to do when the British observatory at Greenwich was the central authority, and most longitudinal measurements were based on estimates of distance from Greenwich. Even the Danish comet medal recognized the centrality of Greenwich, stipulating that the first to discover a comet was the first to report to the observatory in either Britain or Denmark. Britain was the world's astronomical authority; its observatories were richly endowed by the government, which granted the increasingly important professional status to the astronomers who worked there. Cambridge University was doing likewise by establishing a series of chairs and fellowships in astronomy. As the field changed in Europe and England, amateurism became less a badge of nobility and more a sign of incompetence.

When Mitchell discovered the comet and her father reported it to the Bonds at Harvard, the college president at the time, Edward Everett, saw an opening: Mitchell was a remarkably appealing woman whose talent and modesty were equally indisputable. She could never be accused of being a status seeker. But if Everett could convince the Danish government that reporting her discovery to the Harvard Observatory was the equivalent of reporting the discovery to the British Royal Observatory or the Danish Royal Observatory, the Harvard Observatory would gain the status of an international astronomical authority.

Maria was something of a pawn here. She was proud of her discovery, but her intense shyness made her reluctant to publicize it. Yet that shyness was exactly what made her so useful to President Everett. Her friend George Bond had also discovered comets, but he'd been unsuccessful at arguing on his own behalf against the authorities of Europe. Since Bond was directly affiliated with the Harvard College Observatory, Harvard's hands were tied; Everett had never even tried to defend Bond's claims. But by framing Mitchell as something of a damsel in distress, Everett could bring his diplomatic skills to bear on her behalf and establish the precedent that Harvard's observatory was as reliable an authority as the British Royal Observatory at Greenwich.

At the time, Harvard was competing with Cincinnati to establish the observatory that would become the national observatory of record. The seal of approval from the Danish government would greatly help Harvard's effort, and both Mitchells were rooting for the university—partly because William was on the board, but also because by now their friendship with the Bonds had expanded to include a whole circle of Harvard mathematicians and scientists.

On November 10, 1848, Mitchell received notice that the Danish government had awarded her the medal. It was a coup for Harvard as much as for Mitchell, and Everett took care to display the Danish medal to Boston newspapers before he sent it to Nantucket. He wrote to Mitchell, "I have taken the liberty to show it to some friends, such as WC Bond, Professor Pierce, the editors of the 'Transcript,' and members of my family—which I hope you will pardon."[25]

By 1848 Harvard and the U.S. Naval Observatory were working in concert: using a telegraph and a "magnetic" clock, the two observatories developed a new, much more precise method of measuring longitude by calculating the difference between a star's ascension in Cambridge and in Washington. In 1849 they scored another triumph when the U.S. Navy convinced Congress to appropriate money to fund an American nautical almanac, and to locate the project's offices in Cambridge. Harvard had succeeded in establishing itself as the headquarters of American astronomy.

In Washington, D.C., the Smithsonian, the American Academy of Arts and Sciences, and the U.S. Navy were still jockeying to establish themselves as scientific authorities. Eventually, all three institutions would affiliate themselves with Mitchell in one way or another. The first was the American Academy of Arts and Sciences, which elected Mitchell as an honorary member in June 1848—the first woman to be honored in this way. Within months Mitchell had also accepted an appointment as computer for the U.S. Navy's *Nautical Almanac*. One bonus of the almanac job was that she could do it while she continued her work as librarian of the Nantucket Atheneum. Another was that she was invited to spend the summer of 1849 with Bache on the coast of Maine, learning to use the most up-to-date instruments in making observations for the United States Coast Survey. She accepted gladly and retreated for a scientific holiday on the coast of Maine, a welcome respite from the growing public attention. Everybody wanted a piece of her.

There were many international nautical almanacs, but there were two reasons to establish one in the United States. One was simple patriotism. This argument worked well with Congress and the pub-lic, who believed that it would be implicitly good for American mariners and scientists to rely on astronomical predictions made by Americans rather than those made by Europeans. The second reason was purely scientific: the French and British tables that most al-manacs relied on were quite old, in some cases fifty years old. As-tronomy had advanced a great deal in the interim, and American astronomers now had much more accurate observations and more precise methods of calculation to work with. The almanac would be scientifically stronger, giving American astronomers a decided ad-vantage over their European counterparts.

In 1849, Maria Mitchell was invited to join the staff of the new U.S. *Nautical Almanac* as a computer. It was not the top job; the al-manac had a director, Charles Henry Davis, an intellectually ambi-tious naval lieutenant who had been an astronomer before he became an expert navigator. The lead mathematician was Edward Pierce, a professor at Harvard. But being a computer was one of the top jobs, indeed one of the few existing jobs, in astronomy. Simon Newcomb, another computer for the almanac, proudly referred to the group as "an aristocracy of intellect."[26] Mitchell's salary of five hundred dol-lars a year was five times what she had earned at the Nantucket Atheneum, and she could do the work at home or on long, quiet af-ternoons in the library. When she accepted the appointment, Mitchell became, along with her new colleagues, one of the first professional astronomers in the United States.

Like much research in astronomy, Mitchell's work as computer of Venus was primarily mathematical. Using complicated formulas, she reduced prior observations to tables and then made calculations that enabled her to predict Venus's future positions. Such work was useful for navigators, who could locate Venus on any given day and then deduce their own positions. It was also useful for working astronomers, who could rely on the accuracy of the *Nautical Al-manac*'s tables to calibrate their own instruments in order to produce precise results in their own observations. But the work of the al-manac was not observational; it was computational. Mitchell used reams of paper for her calculations, which demanded total accuracy and an immense tolerance for finical, repetitive work. As computer of Venus, Mitchell was at the forefront of the mathematical branch of astronomy, and her discovery of the comet and her work with instruments on the Coast Survey had already placed her among America's most successful observational astronomers. Fortunately, she loved both kinds of work, and her calculations for the almanac proved that she was a tenacious, persistent mathematician as well as an insightful one. She would hold the appointment for nineteen years, until 1868. The great success of the almanac during that time helped establish the credibility and importance of American astronomy to the world at large.

In the years that followed, the honors mounted, though the early honors were the most significant for Mitchell's career. In addition to the medal from Denmark, there was a later medal from the Republic of San Marino. She was granted honorary degrees, memberships in exclusive associations and societies, and remarkable celebrity. In 1849, with Louis Agassiz's sponsorship, she became a member of the American Association for the Advancement of Science (though the association's secretary Asa Grey crossed out "member" on her certificate and penned in "honorary member"). Many years later, in 1869, she was among the first women to be offered membership in the American Philosophical Society. In 1850, she was delighted when her friend Elias Loomis, professor of astronomy at Yale (and one of the men who had helped William and Maria set up their observational platform on the roof of the Pacific Bank), published a textbook on *The Recent Progress of Astronomy* that included an entire chapter on "The Nantucket Comet," the only American astronomical discovery that rated a chapter of its own. That same year Joseph Henry, director of the Smithsonian, finally scraped together the funds to honor her with a one-hundred-dollar cash prize. She became the subject of an endless series of biographical sketches in newspapers and magazines and inspired a procession of tourists to take the ferry out to Nantucket to see her at work in the Atheneum. She found herself imprisoned by her own celebrity, an object of public scrutiny for the rest of her life. But she also found that the little comet would open all sorts of doors for her, creating opportunities she had only dreamed of during those long, quiet nights on the rooftop.

Miss Mitchell's comet was small. Although its discovery and the calculation of its orbit contributed to the great changes in nineteenth-century astronomy, its scientific importance was minor compared to its historical significance. Because of that flash in the night sky, Maria Mitchell became known to her peers as "one of the world's greatest women."[27]

CHAPTER 5

"A Center of Rude Eyes and Tongues"

As she entered her thirties, Mitchell thought a great deal about what her future as scientific woman might be like. Her brothers and sisters all seemed to be marrying, but she doubted that marriage would ever be right for her. Still, she had grown up in a large family and hated the idea of living completely alone. Did her scientific greatness doom her to solitude? Every summer she was crowded by so many tourists and visitors that solitude seemed a pleasant fantasy, but when the dark Nantucket winter came and the tourists disappeared, Mitchell had plenty of time to contemplate her shrinking "home circle." Where would she find love and companionship once her sisters were married?

Winter on Nantucket can be desolate. Summer visitors and many long-term residents leave the island. Most stores and restaurants close, and many people who remain are forced to stop working. The extreme change in the weather makes the ferries unreliable. The winds blow very loudly, sometimes raging for days on end, making it hard to hear anything else. The sky seems to sink into the ocean, the colors seem to disappear, and the gray deepens. Nantucket's November twilight is darkly monochrome, an infinite variety of shades of gray. Gray and white seagulls wing through the dark skies, silver-gray seals float in the gray-green waves, and all the seaweed looks black. The yellows of the sand, the blues of the sea, the greens of the tangled beach roses and grapevines—all subside to gray.

Lonely as it is, winter on Nantucket is starkly beautiful. Although it gets cold, the winters are not as cold as they might be if the island were just a bit farther north—the edge of the Gulf Stream almost brushes the island. Slanting winter sunlight plays over the subtle palette, and the relative quiet, unmarred by the noise of summer traffic, allows the sounds of the ocean to travel across the empty moors. The sky is dramatic, whether cloud fronts are rushing in or wrapping the bushes in fog. Occasionally it is clear, almost crystalline. On these nights, the stars above Nantucket shine with a special brightness that comes from being thirty miles offshore.

In the 1830s and 1840s, as Nantucket started to attract summer tourists, the contrasts between the seasons grew sharper. More and more vacationers came to the island every summer, creating the hectic pace of a beach town. At the same time, winters grew quieter as the whaling business began to drift from Nantucket toward New Bedford. By 1850, Nantucket's economy was certifiably manic-depressive. Every summer there was a temporary boom while every winter the bust got harder and deeper, and the decline of the non-tourist economy became more painfully evident.

For Mitchell personally, the contrast was extreme. In the summer she was busier than ever, particularly because so many tourists wanted to meet her. In the winter, although still employed at the Atheneum, she had more time on her hands than ever. There were few visitors to the library. Many of her younger siblings had left the island, along with several of her childhood friends. She turned to keeping a diary and wrote long, contemplative entries detailing her thoughts about friendship, marriage, and work. (Mitchell may have kept diaries before the 1850s, but none survives. Perhaps she destroyed earlier journals that she had written when she never expected anyone else to read her writings. After the comet, Mitchell seems to have realized that she was writing for posterity.) From about 1850 through 1858, her diaries show that she was attempting to adjust to her new public role, and they tended to be written in the off-season. In the summer she was too busy to think, much less write down her thoughts; virtually every summer tourist stopped into the Atheneum to see her.

During the years that followed her discovery of the comet, countless American newspapers and magazines reported on her whereabouts or offered brief profiles of her. Copyright laws were lax, and rural papers often reprinted articles about her taken from more prominent papers. Today, a quick search of EBSCO's historical newspapers archive between 1847 and 1857 turns up more than three hundred entries on Maria Mitchell, but that search engine is limited to a few papers and magazines of record. The actual number of articles about Mitchell, the "lady astronomer," is impossible to guess. From Mitchell's own perspective, it was a storm of mostly unwanted publicity. She was embarrassed by all of the public attention, and the stream of visitors who started to come to Nantucket in order to get a glimpse of her made her uncomfortable. Some introduced themselves, and a few asked for autographs. Others just gawked, then wrote up their impressions for their hometown newspapers.

Through all these distractions, Mitchell gradually caught up with her father's mathematical and scientific expertise, and then surpassed him. It was private work—self-education, self-improvement, perhaps self-culture—but it was not necessarily directed toward any specific goal. In the 1840s she earned enough money at the Atheneum to be somewhat self-sufficient, but she did not have particular financial needs or goals; her father was earning fifteen times as much as she was, and she shared his free housing. The hours Mitchell spent at the Atheneum in the 1840s and early 1850s were quiet and fundamentally modest. She began to look for a new direction.

When Mitchell started looking for comets, she probably hoped to contribute something to astronomy. She was hunting for an original discovery, and she must have been aware that such an achievement might bring her some public recognition. But the discovery of the comet made her wildly, unexpectedly famous. With her appointment as the computer of Venus, she was officially one of the top astronomers in the United States (indeed, one of the few with a paid position), and her annual earnings increased by 75 percent, making her a highly successful wage earner. What did all of this mean for her? In September 1854, Mitchell wrote:

> I am just through with summer, and summer is always such a trying ordeal. I have determined not to spend so much time at the Atheneum another season. To put someone in my place who shall see all the strange faces and hear all the strange talk... Four women have been delighted to make my acquaintance, three men have thought themselves in the

presence of a superior being, one has offered me twenty five cents because I reached him the key of the museum, one woman has opened a correspondence with me and several have told me that they knew friends of mine. Two have spoken of me in small letters to small newspapers, one said he didn't see me, and one said he did! I have become hardened to all, neither compliment nor quarter dollar rouses any emotion.[1]

Two weeks later, she fielded more offers:

One man and one woman asked me if I was the astronomer, and I do not even change color. One man offered me 50 cents for a sight of the museum and I colored very slightly. One insisted on my giving a sketch of my astronomical labors to a party of his friends and I *did* "color" almost with indignation and declined doing it.[2]

Mitchell's accounts of her own celebrity show her good humor and convey some of her frustration. A year later, her frustration had markedly increased: "Another day of do-nothing," she wrote, not because she had been idle but because a constant stream of visitors had forced her to play hostess, preventing her from reading, calculating, or even thinking. "I am afraid all my days will soon be so and yet it is thus. No fault of mine. It is certain that I must either give up the astronomy altogether or put someone in my place a large part of the year. I was never more dissatisfied with my life than I have been this summer. I wish I knew what was best for me to do."[3] A week later, she added, "I have read almost nothing, except newspapers— how could I? working six hours a day is as much as I can bear, with the exercise and the home duties besides. But it makes me feel miserable to perceive that I am only a money getter."[4] Mitchell's profound dissatisfaction in 1855 came at the end of another long summer, during which the Atheneum had become like a zoo and she had been the captured animal on display.

There is no question that some of these encounters were highly gratifying. Perhaps Mitchell's favorite visitor was the man who had recently published a biography of Margaret Fuller. "Yesterday," Mitchell wrote, "James Freeman Clarke, the biographer of Margaret Fuller, came into the Atheneum. It was plain that he came to see me and not the Institution. I was a good deal embarrassed and made such an effort to appear as if I wasn't, that I was almost ready to burst into a laugh at my own ridiculousness. He rushed into talk at once, mostly on people, and asked me about my astronomical labors. As it was a kind of flattery, I repaid it in kind by asking him about Margaret Fuller."[5] Despite her description, Mitchell's questions about Fuller were not posed out of mere courtesy. Margaret Fuller was a towering figure for intellectual women in the nineteenth century.

In 1839, when Fuller began her "Conversations," private philosophical seminars for women in Boston, Mitchell was twenty-one. Employed at the Atheneum, she didn't have much opportunity to travel to Boston at that time, and her circle of acquaintance was small. Over the next ten years, however, Mitchell had ample opportunity to read Fuller's writings: first the magazine the *Dial,* which Fuller edited and to which she regularly contributed, and then her books *Summer on the Lakes* and *Woman in the Nineteenth Century.* By 1847, when Mitchell discovered the comet and her circle widened, Fuller had left Boston for Italy, where she was caught up in the excitement of the 1848 revolution. During those tumultuous years of Italian unification, Fuller fell in love with a revolutionary count, secretly married him, and had a son. The revolution failed, and in 1850 Fuller left Italy for the United States with her husband and child. Their ship was wrecked off Fire Island, New York, and Margaret Fuller and her family drowned.

It must have been hard for Mitchell to know what conclusions to draw from Fuller's example. Fuller's intellect was truly formidable, and her writing was inspiringly persuasive. But in her greatest work, *Woman in the Nineteenth Century,* published in 1845, Fuller compared marriage to slavery and argued forcefully for women's independence. Within five years of its publication, she was embroiled in a sexual scandal (a secret marriage) and seemed to have sacrificed her interest in women's causes and her own independence in order to nurse the soldiers who fell in the cause of the man she loved. If Margaret Fuller had survived her passage back to America in 1850, she would have faced many questions about her marriage, the legitimacy of her son, and her commitment to her own work. Her tragic death only intensified these questions.

Fuller's life was of particular, even urgent interest to Mitchell, who had few other guides or role models among nineteenth-century

American women. Fuller had even predicted the existence of some-one like Mitchell in 1845, when, in *Woman in the Nineteenth Century,* she laughingly imagined that eventually nature would bring forth a "female Newton." But Fuller's ideas about women in the sciences are difficult to parse. Because she thought of women as having a "superior susceptibility to magnetic or electric influence"—with a special "genius" that is "electrical in movement, intuitive in func-tion, spiritual in tendency"—she didn't think of creating original work or completing projects as particularly feminine.[6] But she did imagine *herself* as someone who did original work and got things done. "[T]here is no wholly masculine man, no purely feminine woman . . . ," Fuller wrote. "Nature provides exception to every rule. She sends women to battle, and sets Hercules spinning; she enables women to bear immense burdens, cold and frost; she enables the man, who feels maternal love, to nourish his infant like a mother. Of late, she plays still gayer pranks. . . . Presently, she will make a female Newton."[7]

As a woman scientist, Mitchell might have been troubled by Fuller's merry, mocking attitude here, but other Fuller writings were pleasingly challenging. Fuller argued for the importance of women's work, a principle dear to Mitchell's self-supporting heart. Most im-portant, Fuller provided a sustained critique of women's confine-ment to the domestic sphere, cautioning women against marriage and arguing that "every path" should be "laid open to woman as freely as to man," and urging women toward "independence of man."[8]

Mitchell's journals show that she thought long and hard about Fuller's ideas, and her own ambitions, during the long, quiet winters between 1850 and 1857. Fuller had written in 1845, "Many women are considering within themselves, what they need that they have not, and what they can have if they find they need it." By 1855, Mitchell was one of these women. In her diary, she wrote that "we were born dependent, and our happiness is in the hands of others,"[9] but she regretted her own dependence:

> The older I grow, the more I admire independence of character and yet
> the less does this characteristic belong to me, and the more rare does it
> seem to be in the world. When we consider, too, how short is life and
> how much shorter are the petty vexations of life, it seems strange that

we should not act up to our convictions of duty and disregard what may be said of us by our fellow men. For what is my neighbor more than I that I should succumb to his view in preference to my own? And what possible good can come to me from such submission? I cannot even please him for very possibly his expressed opinion is not his own but that of some other neighbor of whom he stands in awe.

And so we have a chain of ignoble submission reaching perhaps around the world. I cannot suppose it comes from cowardice and I therefore suppose it comes from a still more despicable weakness—that of indolence. Thinking is hard work.[10]

Mitchell's image of a "chain of ignoble submission" stretching around the world is original and funny, but she doesn't stop there. She traces such submission to indolence and concludes that people are simply too lazy to think for themselves. There is little evidence that Mitchell actually behaved so submissively; instead, she is philosophizing here, trying to articulate her own doctrine of self-reliance.

Mitchell was an industrious thinker during these years. She was on a completely new path: the first woman and the first scientist to be celebrated in the American press to such an extraordinary degree. As John Lankford, a historian of astronomy, puts it, "In the early nineteenth century, writers and intellectuals sought to create heroes to serve as inspirational models for the new American nation." Portraits and profiles of George Washington were everywhere, and stories about his heroic qualities rolled off the presses daily. Other Americans joined the pantheon of heroes, including Lewis and Clark, Andrew Jackson, and even Daniel Webster, but before Maria Mitchell, "there were no heroines or scientists in the national pantheon."[11] As the endless procession of thrill-seeking tourists was teaching her, Mitchell had a demanding public role to fill.

In the 1850s, the roles of scientist and woman were not as contradictory as we might now suppose, but the concept of being a woman and a national hero held enough contradictions to keep anyone pondering through a long winter night. Women were generally seen as creatures of the private sphere. When tourists stared at Mitchell in the Nantucket Atheneum, they were making her public in ways that were quite vulgar, according to the sexual ideology of the time.

The nineteenth-century ideology of separate spheres for women and men held that the public sphere was male and the private sphere was female. Like most social constructs, the concept did not necessarily hold together logically; men spent much of their lives at home, and women spent much of theirs outside the home. But on mid-century Nantucket the separate spheres were more extreme, and perhaps more extremely paradoxical, than they were in most other places. The entire island could be seen as a private sphere, while the oceans (and the world) outside were the public sphere. In that sense, most men did leave the private sphere to live on the ocean for years at a time, often returning for a month or less between voyages of three, four, or even five years. Few women accompanied them off the island, but the stark division between the island and the ocean forced women to carry on much of the town's public business. They managed money; bought, sold, and worked farms; ran businesses; and split their own wood as a matter of course. On the one hand, Nantucket's dependence on whale fishery raised a thick wall between the male and female spheres, but on the other, the absence of the whale men gave many women the opportunity—even the mandate—to enter the spheres of business and civic leadership that in other communities were assigned to men. These paradoxes clearly contributed to the great independence and strength of many Nantucket women. But from the nineteenth century onward, commentators have often remarked that women on Nantucket faced a unique set of challenges. In *Letters from an American Farmer,* published in 1782, Hector St. John de Crèvecoeur, the philosopher and writer, described the dynamic this way: "As the sea excursions are often very long, their wives in their absence are necessarily obliged to transact business, to settle accounts, and in short, to rule and provide for their families. These circumstances, often being repeated, give women the abilities, as well as a taste for that kind of superintendency."[12] While noting the competence of Nantucket wives, Crèvecoeur also noticed that marriage was hard on them. Soon after marriage, he commented, Nantucket women "cease to appear so cheerful and gay."[13] Their husbands were very soon off at sea, and they were left alone with double responsibilities and, not insignificantly, their opium bottles. Crèvecoeur writes:

> A singular custom prevails here among the women, at which I was greatly surprised; and am really at a loss how to account for the original

cause that has introduced in this primitive society so remarkable a fashion, or rather so extraordinary a want. They have adopted these many years the Asiatic custom of taking a dose of opium every morning; and so deeply rooted is it, that they would be at a loss how to live without this indulgence; they would rather be deprived of any necessity than forego their favorite luxury. This is much more prevailing among the women than the men.[14]

It can be hard to reconcile the modest Quakers of our imagination with this group of drug fiends. Historians have long been puzzled by Crèvecoeur's description, since there are no other records of Nantucket's women as opium addicts, but there may be something to the story. The writer Nathaniel Philbrick reports that when sewer workers in the 1980s uncovered some refuse from the great Nantucket fire of 1846, they found a trove of opium bottles and a surprising stash of phallus-like sex toys, many carved from whale bone.[15] Philbrick sees these remains as evidence of the enormous stress these women faced when their husbands went off to sea: "The Nantucket women may uncomplainingly carry on the duties of raising a family without their husbands, but the evidence of stress is undeniable—a stress symptomatic of a society that must tear itself in half (into the two worlds of work and family) if it is to sustain itself economically."[16] When whalers' wives assumed "superintendency" of the family's more public positions, they found themselves increasingly at odds with a larger culture that celebrated women's domesticity. One whaling wife, Sarah Howland, expressed frustration with her inability to make her life conform to her own domestic ideals, declaring plaintively, "I have always had a good home, but I have ever felt homeless."[17] According to the historian Lisa Norling, the contrast between whaling couples' domestic ideals and their realities was heartbreaking. The great emotional hardships faced by maritime wives, she argues, "demonstrate both the tenacious hold of domesticity and its ultimate failure in the maritime context."[18] A feeling of homelessness is surely one of the most constant attendants to the ideology of domesticity. By valorizing the ideal home, domesticity makes everyone who is trapped in a mere good home feel inadequate.

Mitchell was somewhat insulated from the failures of maritime domesticity because her father and mother had an exceptionally close companionate marriage. William Mitchell declined to attend Harvard because he found it impossible to leave Lydia Coleman behind; perhaps for the same reason he never embarked on a whale ship. Maria and her brothers and sisters were among the relatively few Nantucketers whose family lived up to contemporary ideals of domesticity. But Maria knew that her parents' marriage was an exception to the economic rules that governed the island. In the Mitchells' early years, their spheres were hardly separate: Lydia worked as a schoolteacher before William took up teaching. Most of their daughters and a few of their sons would also teach. Scientific work was shared by male and female Mitchells, with, of course, Maria helping her father most frequently, both with astronomical observations and with surveying work.

Later, the whole Mitchell family shared William's position at the bank, where the spheres of home and work were impossible to pull apart, and the spheres of male and female correspondingly hard to define. Another complicating factor in the Mitchells' residence at the bank was that it made them apartment dwellers rather than house dwellers. The urban studies scholar Betsy Klimasmith contends that apartment dwellers tend to feel connected to the public world when they are at home, rather than sheltered or separated from it. In Maria's case, this sense of connection must have been powerful. The Pacific Bank building stands at the top of Main Street. The two or three blocks below it, between the bank and the water, are primarily filled with businesses, which in Mitchell's time included factories and warehouses. Above some of the shops are other apartments, nestled close to one another in the heart of the tiny town. Beyond the bank building, heading west, most of the buildings are grand single-family houses built by the wealthiest residents of the island in the early nineteenth century. The bank is positioned between the business district and the residential district, and also between the modest apartments in the center of town and the fabulous houses just next door. Living above the bank, the Mitchells were linked to some of the poorest inhabitants of the island, who lived in rooms above the nearby shops, but their apartment was palatial.

The quasi-public nature of the Mitchell family home at the Pacific Bank gives particular insight into Mitchell's questions about her own public persona. The little girl growing up on Vestal Street (at the far end of Main Street, where the houses get smaller again) probably had a sense of her home as a private refuge, but the woman who lived upstairs from the commercial and social hub of the town did not have a place to retreat. For the twenty-five years that the Mitchells lived in the bank, nothing was wholly private. Their rooms were a gathering place. Most islanders deposited their money at the Pacific Bank, which made them part shareholders—part owners—of the Mitchell family apartment. A steady stream of visitors poured into the grand front room, with notable off-island guests that included Ralph Waldo Emerson, Herman Melville, Frederick Douglass, and Sojourner Truth. Although she was sometimes reticent in company, Mitchell did not necessarily dislike this social scene; the visiting political, literary, and scientific thinkers whom the Mitchells entertained gave her a sense of connection to the world of ideas. During the 1850s, then, Mitchell's private sphere shrank almost to nonexistence. Her observatory was shared with her father, equipped by the government, and, most important, had become a scientific workplace. Her house doubled as a bank and a social gathering place. Even her diaries had become semipublic documents—family lore holds that she destroyed all of her diaries and private papers shortly after the Great Fire of 1846, when she learned that the strange winds blowing off the fires in many houses had scattered personal papers around the town. From that point on, all of her journals and notebooks were written with an awareness that they might one day be public.

Fuller was one of the few women who published her thoughts about women's place in the public sphere. *Woman in the Nineteenth Century* included a letter describing a controversial public speech against slavery by Angelina Grimké: "The scene was not unheroic— to see that woman, true to humanity and her own nature, a center of rude eyes and tongues, even gentlemen feeling licensed to make part of a species of mob around a female out of her sphere."[19] Mitchell was not yet a public speaker, but she was certainly a "female out of her sphere" who often found herself "a center of rude eyes and tongues." In many ways she was well equipped to deal with this, since the experience of her whaling neighbors thrust most females out of the domestic sphere in one way or another. In addition, she had cousins and old friends, such as the feminist Lucretia Mott and the preacher Phebe Hannaford, who were gaining fame as public speakers, but they were all at odds with her culture. Astronomy was the perfect domestic avocation—a female astronomer who takes her telescope up onto the roof never leaves her family home. The problem, if it was a problem, was Mitchell's growing fame and shrinking sense of privacy.

Mitchell's biggest questions during the 1850s were about work and marriage, and the two were intimately related. For nineteenth-century women, marriage was a career, and for most of Mitchell's peers and all of her sisters, it was what they eventually chose. Many of the women Mitchell knew had worked for a few years as teachers or librarians before marriage (as her mother did), and quite a few had had a temporary interest in science at that time, but most ended up making marriage their ultimate career.

Many of Mitchell's questions about marriage were related to work, but several were also related to the importance of her friendships—her intimate bonds with women. Mitchell's relationships with women, particularly Lizzie Earle and Ida Russell, were among the most important in her life at this time. These intimate friendships strongly influenced her sense of herself as someone connected to an intimate sphere—a person who did not necessarily need to marry to find intimacy.

But the prospect of choosing to be single was a bit daunting. In 1853, when Mitchell went to Boston to attend a scientists' meeting, the reformer Dorothea Dix asked to meet her. Dix was about twenty years older than Mitchell. She was well known for her public campaigns on behalf of the mentally ill, who had been imprisoned in conditions of almost gothic horror before Dix began to speak and write on their behalf. Dix's leadership of the asylum movement placed her in the "firmament of benevolent womanhood," according to *Godey's Lady's Book*.[20] But Mitchell didn't warm to Dix. On the contrary, she commented, "It is sad to see a woman sacrificing the ties of the affections even to do good. I have no doubt Miss Dix does much good, but a woman needs a home and the love of other women at least, if she lives without that of man."[21] Mitchell's perceptions of Dix were accurate: Dix was a solitary figure who felt that she had never found "moral or intellectual companionship" even among those who described themselves as her "warm and true friends."[22] Dix's sense of lonely alienation was frightening to Mitchell.

Throughout the winter of 1853, Maria Mitchell wondered whether she would face the same sort of loneliness if she left Nantucket. Increasingly, she began to fear that she might be lonely even if she stayed. In 1854 her sister Phebe married and moved away, leaving a huge gap in Mitchell's "home circle," as she called it. She was despondent: "I have had no sickness at heart for a long time comparable to that which Phebe's absence gives me. I could cry daily at the things for which I miss her. She had so much mind and was almost always with me good natured and that is invaluable in the home circle. When you are seasick and a vessel goes madly up on a wave and plunges recklessly down into the gulf below, how she shakes her sides and seems to be convulsed as if with a chuckle at your heaving indignation. What a dreary desolate feeling it gives you, as if your only friend had deserted you..."[23] On mournful winter mornings in the Atheneum, Mitchell thought about striking out on her own, as the passionate Margaret Fuller and the lonely Dorothea Dix had done. She reported that "Mr. —— somewhat ridicules my plan of reading Milton with a view to his astronomy, but I have found it very pleasant, and I have certainly a juster idea of Milton's greatness than I had before. I have filled several sheets with my annotations on the 'Paradise Lost,' which I may find useful if I should ever be obliged to teach, either as a schoolma'am or a lecturer."[24] Mitchell knew she was unlikely to be obliged to teach, but the idea of being a lecturer was appealing. As librarian of the Atheneum and as a long-standing member of the Nantucket Philosophical Institution (and the daughter of William Mitchell), she was intimately familiar with the art of the thoughtful lecture. The social strictures against women speaking in public were intense at the time, but they probably seemed less forbidding to Mitchell than they would have if she had not already found herself "a center of rude eyes and tongues."[25]

Still, Maria Mitchell was shy. It was a struggle for her to keep her countenance and control her blushes when strangers expressed interest in her. Despite all of her work organizing public lectures, she feared climbing onto a speaker's platform herself. Another factor holding her back from becoming a public speaker (or even a publishing scholar) may have been her fear of solitude: she certainly did not want to end up like Dorothea Dix, distanced from her friends and from her family. But perhaps the most important muzzle on Mitchell's public persona was her love for Ida Russell.

There is no record of how Maria Mitchell met Ida Russell or what cemented the women's friendship. Russell was not a Quaker, a Nantucketer, or an astronomer; she was simply an intellectual. In the years before women were formally educated, when few intellectual women published their ideas or spoke publicly about their accomplishments, meeting a kindred spirit was a lucky break. Mitchell and Russell were the same age, and they moved in Boston's most prominent intellectual circles.

Ida Russell was born in Sweden in 1818. Her father, Jonathan Russell, graduated from Brown University and became a lawyer and diplomat; he served in England during the War of 1812 and as ambassador to Norway and Sweden in 1818. In 1839, shortly after they moved to Massachusetts, Ida and her mother attended the first year of Margaret Fuller's "Conversations." Sophia Peabody, who would eventually marry Nathaniel Hawthorne, also attended Fuller's "Conversations," and Ida and Sophia became friends. Ida lived in Milton, Massachusetts, at that time, in the grand Hutchinson mansion.[26] Nathaniel Hawthorne was a little jealous of Sophia's friendship with Ida; on November 27, 1841, he wrote a passionate love letter to Sophia, begging her to meet him and exclaiming, "How could I have borne it, if thy visit to Ida Russell were to commence before my return to thine arms!"[27]

Nathaniel Hawthorne was not the only one who was jealous of Ida. She was as beautiful as she was intelligent, and since she never married, many prominent men were rumored to have fallen in love with her. In 1842, Ida Russell and her half sister Amelia moved to Brook Farm, the utopian community west of Boston that Nathaniel Hawthorne had recently quit. Ida spent the summers there in 1842, 1843, and 1844, and her sister wrote a memoir of Brook Farm that was published in the *Atlantic Monthly* in 1878.[28] Orestes Brownson Jr., son of the philosopher Orestes Brownson, was one of Ida's Brook Farm suitors. Apparently Ida Russell "aroused in him a passion for the sea by singing 'A Life on the Ocean Wave,'" and the fourteen-year-old Brownson quit Brook Farm and went away to sea.[29] Later, the Catholic theologian Isaac Hecker fell in love with Ida Russell at Brook Farm, although he seems to have fallen in love with most of the women residents.[30]

By 1846, Brook Farm was far behind Ida, and she seems to have come close to marrying one of the distinguished men who pursued her: John Greenleaf Whittier. Sarah Helen Whitman, the popular Providence poet, had probably heard about Russell's purported engagement but she dismissed the possibility. Whitman, who thought of Ida Russell as "one of my dearest friends," wrote that Ida Russell "was a splendid, beautiful creature, but lymphatic and impeded by worldly cares and solicitudes and ambitions. [Whittier's] Quaker simplicity and her bumptious tastes did not readily harmonize, so they grew to be strangers."[31]

No letters between Mitchell and Russell survive, but we can surmise their intellectual level by reading the letters from Sarah Helen Whitman to Ida Russell. Whitman wrote to Russell of her own "profound plunge into Hegelian philosophy"[32] and of Edgar Allan Poe's review of Elizabeth Barrett Browning's poetry.[33] Mitchell and Russell's own correspondence was probably equally stimulating.

Such intellectual ardor found particularly fertile ground in New England. The hostess and New York salon leader Mrs. Ripley, who had once been a Brook Farmer, explained to Mitchell that "in Society such conversation as you have had with Ida Russell is unknown, books are not mentioned, the words you use would not be understood. In New England, women talk of the Fugitive Slave Laws. In New York society, it is never talked of, and downtown even, it is talked of by very few men. I have met with the finest minds in New York, but they are not in Society."[34] For Mrs. Ripley, Mitchell and Russell were among the finest minds of their generation.

During these years, Mitchell's friendships were increasingly important to her. At times she worried about her dependence on friends, while at others, she marveled at her friends' love for her. Mitchell wrote most about her friendship with Ida Russell: "Ida is always very *expressive* to me—she seems really to be much attached to me and I suppose it is the evil of my own nature which makes me ask 'Can it be real' but I hope it is only that setting so high a value as I do upon anyone's affections and giving it so charily to others as I do myself, I cannot believe that I have really won so much of anyone's heart."[35] It was a great, unsettling honor to win the heart of a woman such as Ida Russell, and Mitchell's deep friendship for Ida was complicated by Ida's outright disapproval of public women. Mitchell wrote about Ida's disgust in her diary:

> Ida is strong in her dislike of the "platform women" as she calls the Antislavery and woman's rights people. I told her not to speak of them with such contempt as I had always felt that when I was pushed for money I could write some astronomical lectures and go into the cities and deliver them. "Don't Maria," she said, "do anything else. Take a husband even!"
>
> (I) What! The weak minded man such as would happen to fall in love with me?
>
> (Ida) Well, then there's the river.
>
> (I) Yes, as the bowl or the dagger but I don't fancy any of these things![36]

This awkward conversation presents the crux of the problem. At this point in her life, Mitchell's only real choices were the marriage bed or the public platform, and her dearest friend was adamantly opposed to the latter. The two of them laughed about it—Ida was the first to suggest that Mitchell should just drown herself—but the problem was real. Maria Mitchell and Ida Russell were caught up in the passions of a mid-nineteenth-century intimate friendship. Their relationship was hard for them to understand, and it is even harder for us to comprehend 150 years later. Some of Mitchell's chroniclers have tried to defend her from the charges of lesbianism—the "Vassar libel," as it would come to be known—while others have sought to claim her as a poster girl for nineteenth-century Sapphic bliss. The truth is inaccessible to us, and at least in the case of Mitchell's relationship to Ida Russell, it seems to have been fairly inaccessible to her, too. Mitchell wrote, "I have certainly some female friends who are strongly attached to me and the longer I live the more do I value the love of my own sex, which is not seeking for gratified vanity but affection. And yet I should hate to think that men are incapable of disinterested regard!"[37] She clearly valued her love for women and theirs for her, but she understood it as "disinterested," which may well have meant dispassionate or nonsexual. Elaborating on her feelings for two different friends, she wrote,

> Last night I had two letters which did me good. One was from Lizzie Earle and one from Ida Russell. The love of one's own sex is precious for it is neither provoked by vanity nor retained by flattery; it is genuine and sincere. I am grateful that I have had much of this in my life. I am sometimes sorry that those who give me so much, should give it to me when it might be so well suited to the domestic station of a wife and I am humbled when I consider that they give it to me because they know me so little—that, living in the same town with me, they would know me better and love me less. I have an entirely different regard for Lizzie and for Ida. I love Lizzie as one loves a sister, I admire Ida and am jealous of her regard for others. It is something like *love* and less generous than that which I have for Lizzie, which is affection. But all these affections are weak compared with what one has for kindred. The ties of blood are stronger than all other ties. Strangely enough, I dreamed of Ida a few nights since. I had tho't so much of what has been said of her in connection with Whittier that I dreamed I saw them together. I wonder if it was a glimpse of a scene in a future state. The parlor in which I seemed to see them was the dirty one of Odeon House, and anything but ethereal.[38]

Mitchell seems a little jealous to think that Ida might marry Whittier. But if Mitchell's love for Ida Russell did tend toward erotic passion, Mitchell did not seem to want to pursue it. She chided herself:

> A friend is not to be found in the world such as one can conceive of, such as one needs, for no human being unites so many of the attributes of God as we feel our nature requires in one who shall be guide, counselor, well-wisher and the like. We have therefore a circle whom we call friends, giving a name to the whole, which perhaps in its singular occupation might be used for the combination. Out of the whole circle we may make up a single friend. We love them all but we love the union of all better. From this one, we have the intellectual stimulus of his higher nature. We become active as we see him active, we gird ourselves for labor as we see him in the struggle, we are more of a man because we know one. From another we have the warmth of affection and our hearts grow as if in a summer feeling. We are ordered in good work, we

are sympathizing to all mankind, we look at the ones of our race with a more lenient eye, for we see with the eyes of our friend. So a third neither stimulates the intellect nor warms the heart, but he cheers us. We are more elastic and buoyant, more happy and radiating more happiness because we know him. They are the jolly men of the earth and may earth be jolly to them, for we could not spare them. Whatever our degree of friends may be, we come more under their influence than we are aware.[39]

Determined to build a large circle of friends and not give herself exclusively to Ida, Mitchell started 1855 with a resolution to be more independent. And then in February came the terrible letter. In her journal, Mitchell wrote, "As our circle of friends narrows, they naturally seem to clasp us in a closer embrace. It is the sad mercy of growing old, that we outlive one and another of those we love. I heard of Ida's death today. It is a blow, sudden and severe."[40]

Mitchell's loss was incalculable, and she grieved for years. In her diary she pinned something, since removed, most probably the announcement of Ida's funeral. Above it, she wrote, "Ida was my superior," and along the side she added, "She was many sided but crystal-like. Every side gave a gleam of light."[41] Mitchell's diary has been carefully preserved in the archives of the Mitchell association. The blank place where the lost memento was once pinned offers a chilling illustration of Mitchell's feeling of emptiness; two other friends died the same winter, and Mitchell's mother fell very ill. But although the loss was hard to bear, Mitchell found her circumstances at the time almost equally impossible. Writing of Ida's funeral, she commented, "I find it harder than such a pilgrimage to bear the crude manners of the people who come to the Atheneum to kill time."[42] With Ida gone, Mitchell could now give serious consideration to the life of a "platform woman" that her friend had so strenuously opposed.

For years, women's rights advocates, American scientists, and the popular press had urged Mitchell to go public. In 1857, an editorial in the *United States* magazine declared, "Miss Mitchell is, like most of our distinguished women, better known and better appreciated abroad than at home. She lives in great seclusion upon the barren island of Nantucket, the very place to nurse grand ideas and promote solitary star-gazing . . . We wish, for the sake of her sex, and from the love of promoting science therein, Maria Mitchell would be induced to exhibit her fine talents in a lecture room."[43] Mitchell would never have been tempted to make a spectacle of herself for her own sake, but for the sake of her sex and for science, it was hard to resist the calls. An article in *Godey's Lady's Book* explained, "Nantucket raises neither fruits nor vegetables for the foreign market, but its supply of brains seems to bear almost any pressure of exportation, and men and women of the grandest type, from positions of trust and honor all over the land, look back with loving pride to the sea-girt isle that sheltered their infancy and youth. Among these, no name is more widely known and respected than that of Maria Mitchell."[44] Mitchell's renown was based on her work; she hadn't necessarily sought fame, but she began to consider using it to aid the causes she believed in.

While Ida had been her dearest friend, Mitchell had gone along with her advice, continuing to work quietly at the Atheneum and to moonlight for the *Nautical Almanac* as the computer of Venus. She had tried to keep a low profile, but she had documented her ongoing computations, theories, and ideas for possible lectures should the opportunity ever arise. After Ida died, there was no longer any compelling reason to stay sequestered on Nantucket. Mitchell decided she was ready to begin living on the public stage. She made preparations to resign from the Atheneum and travel. Mitchell was thirty-eight years old, and she was ready to leave home.

The Shoulders of Giants

Americans who love English literature feel strangely at home in England because they have spent so many hours imagining it. Thanks to Charles Dickens and Anthony Trollope, George Eliot and Charlotte Brontë, and especially to the loving adaptations of their novels served up by the BBC and *Masterpiece Theater,* Victorian London and the hedgerowed lanes of the English countryside feel familiar to us even if we haven't actually seen them. For Maria Mitchell, of course, Dickensian England wasn't the past—it was a far-off present, on a culturally dominant continent that she had often imagined. Maria Mitchell could easily recognize the landmarks of London, Paris, and Rome because she had seen so many engravings of them, along with countless pictures of the English countryside and the Tuscan hills. She had grown up with England and Italy in her own home as her mother read aloud by the fire from Maria Edgeworth or Charles Dickens or as she herself pored over Lord Byron's *Childe Harold* or Elizabeth Barrett Browning's *Aurora Leigh*.

Mitchell would actually see Charles Dickens while she was in London, when she attended a public reading of *The Cricket on the Hearth,* but long before that, his books had made her feel that she'd seen the city through his eyes. In 1846, Mitchell wrote:

> What though to me it is denied
> To roam o'er countries far and wide
> By bard and poet sung
> Give me but books. In thought I stroll

To every land from pole to pole
And talk in every tongue.[1]

Having strolled around London and the rest of Europe in her thoughts, Mitchell thought of herself as cosmopolitan, but she knew she would never truly earn that distinction unless she traveled in actuality.

In 1848, Margaret Fuller explained the nineteenth-century American's need to go to Europe in terms of cultural self-discovery:

> The American in Europe, if a thinking mind, can only become more American. In some respects it is a great pleasure to be here. Although we have an independent political existence, our position toward Europe, as to literature and the arts, is still that of a colony, and one feels the same joy here that is experienced by the colonist returning to the parent home. What was but a picture to us becomes reality, remote allusions and derivations trouble no more: we see the pattern of the stuff and understand the whole tapestry.[2]

In order to "understand the whole tapestry," Mitchell needed to visit Europe—and so, in the fall of 1856, she resigned from the Atheneum.

It was time for Mitchell to meet some of the figures who had loomed large in her imagined world. She wanted to see Dickens, of course, because her mother insisted that each of his novels be read aloud in the Mitchells' parlor as soon as it was published, and she longed to be introduced to Elizabeth Barrett Browning, since she had urged every neighbor on Nantucket to read *Aurora Leigh,* Browning's epic poem about a woman of genius. Mitchell also declared that she was determined to step in the footprints of Newton, Shakespeare, Milton, and Johnson. But the figures who loomed largest in her imagination were the great nineteenth-century scientists. Before she left, Mitchell wrote to Benjamin Silliman at Yale, to William and George Bond and Edward Everett at Harvard, and to Joseph Henry at the Smithsonian asking for letters of introduction. In England, she particularly hoped to meet Sir George Airy, John Herschel, and William Whewell; in France, she hoped to meet Urbain Leverrier. She planned her itinerary in Italy around the schedules of Father Angelo Secchi in Rome and Mary Somerville in Florence. Her purpose in continuing on to Vienna and Berlin was to try to meet Johann Franz Encke and Alexander von Humboldt. Originally she planned to continue on to Russia to meet Wilhelm Struve, the director of the Pulkova Observatory, but she happened to meet him in England instead.

Mitchell approached her trip to Europe as a professional, not a vacationer. She wanted to know what was happening in European science. She had already decided that upon her return, she would not go back to the Atheneum but would endeavor to devote herself to astronomical research, relying for sustenance on her salary from the *Nautical Almanac.* But what should she research, and how? What equipment was necessary for a state-of-the-art observatory, and how should she procure it? Once she equipped herself, where should she point her telescopes? Mitchell wanted to know what the central questions in astronomy in the 1850s actually were, and she wanted to know how European astronomers were approaching them. When Mitchell received her letters of introduction, she made an implicit promise to inform her colleagues of the state of European astronomy upon her return.

In 1857 there were a few scientific journals in Europe and America, but scientific publishing was relatively slow. Many scientists published comprehensive books rather than targeted articles, and researchers waited, sometimes for years, until their results were confirmed before they published at all. Without a network of universities, conferences, and professional associations, it was hard to know what other scientists were working on, rather than what they had already finished. But Mitchell and her colleagues at Harvard, Yale, and the Smithsonian needed to know the big questions in astronomy if they were to have any hope of answering them. One of the most important steps in becoming a researcher is determining what your fellow researchers *don't* know. Once you identify an important open question, you are well on your way to making an important contribution. Discovering new objects in the sky, such as comets or even planets, is one way to do this, but Mitchell wanted a research program with more promising methods than simply watching, waiting, and hoping to be first.

Mitchell's feelings toward her journey were mixed. She was determined to introduce herself to the leading astronomers in the world in order to build her own scientific career, but she dreaded the prospect of traveling alone. She was confident in her work but much less sure of her ability to navigate coaches and carriages, trains and steamers. Eventually she found Prudence Swift, a wealthy young woman who wanted a traveling companion. In exchange for Mitchell's supervision of his daughter, Prudie's father would pay both women's expenses. Mitchell did not need the financial support; she had saved quite a bit of money over the years, and she could continue to draw her salary because she would do computations for the *Nautical Almanac* while she traveled. But she did need company, and being a young girl's chaperone suited her Yankee thrift: she could actually work two jobs (companion and computer of Venus) while she made the Grand Tour. Before they left for England, Mitchell and Swift set out together on a trial journey around the United States. They met in St. Louis and then traveled by paddle-wheel steamer down the Mississippi to New Orleans. There, Mitchell visited the slave markets and talked to many slaves, slave owners, and even a slave trader. She struggled to maintain her composure in the face of what she saw as a grave injustice. Her experiences in the American South on the eve of the Civil War made her terribly uncomfortable; she was not at all at home among slaveholding southerners, and they disliked her in turn because her Quaker style of dress identified her as antislavery before she even opened her mouth. By April, Mitchell was thoroughly convinced that the Union could not last.[3] Later, as the Civil War raged, Mitchell would spend years thinking over her southern trip, but at this time, Mitchell was much more interested in looking ahead to Europe. The American South felt like a foreign country to her. She hoped to feel much more at home in the scientific salons of England, France, and Italy, a hope that would be roundly satisfied.

Back on Nantucket, Mitchell spent a few weeks getting ready for her Atlantic crossing, and on July 22, 1857, she and Swift boarded the steamship *Arabia* to sail from New York to Liverpool. The weather was good, and aside from a near collision with another ship (which the passengers only heard about long afterward), the sailing was uneventful. After ten days on board, the pair disembarked at Liverpool, the largest port in the world. As a Nantucketer, Mitchell was normally at home on the docks; she had spent most of her life within a few blocks of the wharves of Nantucket, but Nantucket was a shadow of Liverpool. At the height of its whaling glory in the early 1840s, Nantucket had boasted ninety ships, and generally there were no more than half a dozen in its port at any given time. At Liverpool's own height, in 1860, just three years after Mitchell docked there, the port cleared 10,260 ships in a single year, with hundreds of ships in port on almost any day. It was busier than any port Mitchell had ever seen; thousands of masts bobbed almost parallel to one another, while the stink of tar and fish and coastal sludge was layered with the new smells of coal, smoke, and steam from the steamers. Liverpool was a sooty chaos.

Perhaps Mitchell steeled herself for the experience of landing at Liverpool by reading Herman Melville's 1850 novel *Redburn*. The full title of the novel is *Redburn: His First Voyage, Being the Sailor-Boy Confessions and Reminiscences of the Son-of-a-Gentleman, in the Merchant Service*. It was the sort of book that would have flown off the shelves of the Atheneum, passed from boy to boy and equally demanded by restless women and girls whose curiosity was less likely to be satisfied by actual voyages. Whether Mitchell had read it or not, Melville's description of Liverpool tells us what she saw.

> Nothing can exceed the bustle and activity displayed along these quays during the day; bales, crates, boxes and cases are being tumbled about by thousands of laborers; trucks are coming and going; dockmasters are shouting; sailors of all nations are singing out at their ropes; and all this commotion is greatly increased by the resounding from the lofty walls that hem in the din.
>
> ... Here are brought together the remotest limits of the earth; and in the collective spars and timbers of these ships, all the forests of the globe are represented, as in a grand parliament of masts. Canada and New Zealand send their pines; America her live oak; India her teak; Norway her spruce; and the Right Honorable Mahogany, member for Honduras and Campeachy, is soon at his post by the wheel. Here, under the beneficent sway of the Genius of Commerce, all climes and countries embrace; and yardarm touches yardarm in brotherly love.[4]

But although Melville celebrated the joyful, cosmopolitan cacophony of the docks, he found the city itself quite disturbing. One complaint he made was that everything was new. Liverpool was a thoroughly nineteenth-century city, with few relics of the preindustrial era. For an American who expected to see the Old World, the newness of Liverpool was disappointing. The industrial economy was also quite shocking—the soot and smoke were overwhelming, while the poverty could be cruel and deadly. Displaced people, unmoored from their village communities, drifted hopelessly around the industrial city; they had nowhere to shelter and nothing to do except wait for death by starvation or exposure. *Redburn* narrates an excruciating unsuccessful effort to find help for a starving woman and her daughters. Ironically, the city was well prepared to deal with the corpses of the woman and her children—the moment they were actually dead, the system swung into action and their bodies were collected and brought inside to the Dead House reserved for starved paupers. Melville's fictional picture of Liverpool's poverty is historically accurate. The book cries out against the injustice of the modern city, but Melville seems nearly as upset by Liverpool's modernity.

Mitchell may have shared Redburn's disappointment at Liverpool's newness; not until she got to Manchester would she experience a European art museum. But Manchester, too, was an industrial city. It was not until ten days after she had landed in England, and four days after she had left Liverpool, that Mitchell saw something that seemed appropriately historic—Worcester Cathedral. "For the first time," she wrote, "I felt that I was in a land which had a 'past,' for I saw a Cathedral."[5]

Of course, the primary purpose of Mitchell's journey was to discern the future of scientific research, so although she appreciated the past, she was not necessarily outraged by innovation. Liverpool was the first stop on her scientific tour, and she met three astronomers there: John Taylor, the retired founder of the Liverpool observatory; John Hartnup, its director; and William Lassell, a prosperous Liverpool brewer and telescope aficionado who built sophisticated reflecting telescopes and used them to search for moons around distant planets. All of the telescopes in Liverpool were impressive: Lassell's private observatory and the Liverpool observatory both had equatorially mounted telescopes that could be precisely aligned with the earth's axis, and all were constructed with careful ingenuity. Mitchell was impressed that Taylor had "invented a little machine, for showing the approximate position of a comet, given its elements,"[6] and she liked Lassell even better, commenting that "Mr. Lassell must have been a most indefatigable worker as well as a most ingenious man; for besides constructing his own instruments, he has found time to make discoveries. He is besides, very genial and pleasant."[7] Lassell had built two large Newtonian telescopes for himself, one ten feet long, the other twenty. Mitchell took careful notes on how he had set up his equipment, from the angle of his eyepieces to his somewhat precarious method of hanging from a ladder suspended from the roof of his observatory dome when he wanted to make high-altitude observations. But she also commented with some asperity, "Mr. Lassell works only for his own amusement" and "Neither Mr. Lassell nor Mr. Hartnup makes regular observations."[8]

If Mitchell was searching for a community of pure scientists who could offer new directions for her research, she was disappointed in her first encounters in Liverpool. Taylor was retired, Lassell was an inspired amateur of great wealth, and Hartnup was a civil servant whose observatory was devoted strictly to the needs of commerce. The main purpose of the Liverpool observatory had been to establish its own longitude in order to set the clocks for the British fleet. Maria and William Mitchell had already done the same thing in Nantucket, establishing Nantucket's meridian and recalibrating dozens of chronometers for captains sailing out of Nantucket. Lassell was also an all-too-familiar type: a gifted amateur with a knack for discovering small celestial objects, just as Mitchell herself had been when she discovered the comet. This was precisely the role that she wanted to move away from.

One reason that neglected telescopes made Mitchell so uncomfortable was that she didn't own a good telescope herself; the little Dolland she had used to discover the comet belonged to her father. It must have been terribly frustrating to see great telescopes that were barely being used. Later on, after she had gotten to know the formidable George Airy, the astronomer who directed the Greenwich Observatory (and who was, therefore, the "Astronomer Royal"), she argued with him over this issue. Expressing his frustration that people were always more eager to invest in telescopes than to employ astronomers, Airy declared that "he would gladly destroy one-half of the meridian instruments of the world, by way of reform." Tartly, the underequipped Mitchell "told him that my reform movement would be to bring together the astronomers who had no instruments and the instruments which had no astronomers."[9]

Despite her lack of professional progress, Mitchell seems to have enjoyed her first few days in England. She was surprised at the cordial welcome she received not just from astronomers but also from other cultural leaders. The great Unitarian minister James Martineau called on her within two days of her arrival, and the American consul at Liverpool, the novelist Nathaniel Hawthorne, followed a day later. On August 4, Mitchell wrote, "Mr. Martineau is one of the handsomest men I ever saw."[10] On August 5, the famously beautiful Hawthorne sat down in her parlor, but Mitchell was not impressed. "He is not at all handsome," she wrote, "but looks as the author of his works should look: a little strange and odd, as if not quite of this earth."[11] Martineau must have been fetching indeed, for Hawthorne was generally thought to be "Handsomer than Lord Byron!" as Elizabeth Peabody, Mitchell's friend and Hawthorne's sister-in-law, had put it.[12] In any case, Mitchell was delighted to find that her reputation had preceded her and that she would receive a more generous welcome than she had dared hope for.

Mitchell set out for London with unexpected confidence. She was mastering the logistics of travel and finding the sights exciting and the society welcoming, but the environment was still occasionally overwhelming. Liverpool's docks were hundreds of times larger and busier than any port she had seen before, but the city itself was manageable; however, London was the largest city in the world. Nantucket had a population of about 9,000 people; Boston had around 137,000; New York had 696,000; and London had more than 2,362,000. In Nantucket town, Mitchell had grown up with a relatively high population density of about 200 people per square mile. London in the 1850s was a hundred times more crowded, with approximately 20,000 people crammed into each square mile. Although Mitchell had seen many engravings and illustrations of the London cityscape and might have visualized it when she read Dickens's novels, nothing could have prepared her for the chaos and the energy of the Victorian city. On August 14, Mitchell wrote, "Today we took a brougham and drove around for hours. Of course we didn't see London, and if we stay a month we shall still know nothing of it, it is so immense.... I never saw such narrow streets, even in Boston."[13] London was full of paradoxes—immensity and narrowness, grandeur and poverty, all coated in smoke so thick that Dickens described it as "a soft black drizzle, with flakes of soot in it as big as full-grown snowflakes."[14] For an astronomer, the smog and soot of London must have been upsetting—there was not much chance of seeing the stars. But the history was thrilling: Mitchell was curious about Samuel Johnson, John Milton, and William Shakespeare, and she wanted to reel back through the centuries and try to find traces of the London known to each of these men. Most of all, she wanted to learn everything she could about her astronomical heroes Isaac Newton and the Herschels.

Isaac Newton was already an established scientist when he moved to the Tower of London in 1696 to take charge of the mint. Newton's predecessors had treated the post as a sinecure, but he was fascinated by things chemical and alchemical, including molten metal, so Newton was an active leader of the mint for thirty years. During this time he remained deeply involved with mathematical and scientific research, publishing multiple editions of *Principia* and *Opticks* along with a host of essays and other books, and assuming the presidency of the Royal Society from 1703 to his death in 1727. Under his direction, the Royal Society investigated the strange coincidence of Newton and Gottfried Leibniz's simultaneous discoveries of integral calculus; unfairly, British scientists accused the German mathematician of plagiarizing from the Briton. Newton's pettiness toward Leibniz is all the more disappointing because Newton's accomplishments in optics and astronomy were at least as significant as his mathematical innovations. Newton's great contribution to astronomy, and to physics, was not merely his development of calculus; in fact, Leibniz's techniques were a bit more graceful and efficient. It was what Newton did with calculus that was transformative. Applying new mathematical techniques to the physical problems of orbits, Newton moved astronomy from pure geometry toward something numerically calculable. Newton was one of the first in a long line of unifiers (like Maxwell and Einstein) who were able to make the physical world numerically intelligible with such accurate grace that mathematical formulations could be used to predict physical realities. But although Newton fought hard against Leibniz to claim the credit for developing calculus, he was also well known for modestly declaring, "If I have seen a little farther than others, it was because I stood on the shoulders of giants." Perhaps just as important as Newton's great work as an astrophysicist (long before that term was coined) was his effort to create a scientific community in which scientists could build on one another's results rather than work in solitude. In the late seventeenth century, Newton was one of the earliest practitioners of a Baconian model of scientific collaboration. Like most astronomers of her day, Mitchell thought of Newton as one of the founders of modern astronomy.

Her first stop in London was Newton's grave in Westminster Abbey. She examined his sarcophagus very closely and later described it in detail in her journal. She thought his head was very fine, and she particularly admired the figure of Urania seated on a globe that he gestured toward. After reverently studying his sarcophagus, Mitchell stepped back, looked down, and realized that she had been standing on Newton himself—rather than being interred in the ornamental sarcophagus, the great philosopher's remains lay beneath a simple slab underfoot. Mitchell must have been taken aback to find herself literally standing on the shoulders of a giant.

Mitchell then proceeded to the British Museum, where she asked to see everything connected with Newton. The curators had little to offer, but they did show her a letter from Newton to Leibniz written in Latin and interspersed with geometric diagrams. She also got a glimpse of a beautifully illustrated manuscript copybook made by Queen Elizabeth. It took some determined sleuth work, however, to locate Isaac Newton's London observatory. Mitchell had heard that he lived in the neighborhood known as St. Martin's, so she hired a horse-drawn cab to take her there. The driver dropped her off at the Newton Hotel. Stepping boldly into a wine shop, Mitchell asked the woman pouring drams at the bar if she knew where Isaac Newton had lived and worked. The bartender stepped into the street and pointed to the top of a small building beside the church where there was a "little, oblong-shaped observatory, built apparently of wood and blackened by age" huddled among the chimneys.[15] There was no way to get up to the rooftop, so Mitchell contemplated the small, seemingly improvised structure from the street. Her sense of connection to the world's astronomers was building and deepening.

On September 10, Mitchell visited Burlington House, the new headquarters of the Royal Society, over which Newton had presided in the early eighteenth century. She asked to see the reflecting telescope that Newton had invented, constructed, and used for years. The telescope was not mounted and she could not take it outside, but she held it in her hands.

London was a hive of scientific activity in the 1850s. At both the Royal Society and at Cambridge, where Mitchell would later tour every inch of Newton's college, the living scientists were more interesting than the historical ones. She was entitled to her welcome at the Royal Society; she had published two notices of her comet in their publications. But she probably felt far more humble than entitled; the men who gathered at Burlington House were the greatest scientists of the age, and all were conscious that the work they were doing was changing science immeasurably. Of course, there were the astronomers—George Airy at Greenwich, John Herschel at Collingwood, in Kent, and John Couch Adams at Cambridge—but this was a golden age for British science in general. The members of the Lunar Society had given the scientific community a jolt at the end of the eighteenth century when they leaped over social and educational boundaries to start an alternative club that competed with the Royal Society and challenged it to greater heights. By the middle of the nineteenth century, the Royal Society had lost much of its snobbery, but Michael Faraday, the great experimenter who definitively linked electricity and magnetism and conceived of the electric motor along the way, always felt more at home at the Royal Institution, whose focus on applied science made it more welcoming to scientists with working-class or industrial backgrounds. In 1856, James Clerke Maxwell took Faraday's work one step further by publishing the first of his papers on electromagnetism and was well on his way to offering mathematical explanations for electromagnetic phenomena. The 1850s were equally thrilling for geologists and chemists, who thronged the Royal Society to hear lectures by the geologist Charles Lyell, who spoke about everything from volcanoes to dinosaurs as he expounded on his theory of the gradual changing of the earth. Chemists were fascinated by August Wilhelm Von Hoffmann's models showing molecular structure, which looked much like the orreries that astronomers had long used to explain planetary orbits. As individual scientists focused on smaller and more precise questions, their fields were merging and complementing one another in ways unimaginable a century earlier. Mitchell visited London shortly before Charles Darwin published *The Origin of Species,* but she still got a taste of the crackling scientific excitement that fostered both Darwin and Hoffmann, and she was welcomed into their circles.

Britain's distinguished scientists liked Maria Mitchell. Some might have been predisposed to like her because she carried glowing letters of introduction, but most were probably curious about the American lady astronomer. With one notable exception, virtually everyone she met became an ally, and many welcomed her into their families. When she arrived in London, she was greeted by a cordial note from Annarella Smyth, whose husband, Admiral William Henry Smyth, was the author of Maria's and her father's well-thumbed book on double stars. Admiral Smyth had advised that the comet medal be awarded to her. His wife wrote, "Admiral Smyth joins me in rejoicing to learn that you are so near and we hope you will soon gratify us with a visit."[16] George Airy and his wife, Richarda, were equally welcoming to her, in spite of Airy's reputation for grouchiness, and Richarda became her best guide through the twists and turns of Britain's scientific society. Years afterward, Mitchell told this humorous story of their meeting:

> In visiting Europe some years since with the definite purpose of traveling for study, I accepted whatever letters were offered to aid me in my efforts. Among others, one of my scientific friends sent me half a dozen letters of introduction, and then in a private note said, "I dare not give you a letter to the 'Bear of Blackheath.'" Many times while crossing the Atlantic, I wondered who the Bear of Blackheath might be. One of the first friends I made in London was Mr. Airy, the astronomer royal at Greenwich. I was adopted at once as one of the household, and upon the care of that family my comfort in the whole of my tour largely depended. But sitting one day in the drawing-room with the astronomer royal, I looked out upon the beautiful country around and asked, "What is this charming region called?" He replied, "Blackheath," and I woke to the realization that I was talking with the "Bear."

Perhaps the reason George Airy grew so fond of Mitchell was that on their first meeting she showed him something that had never been seen in England before: a photograph of the stars made by the photographer John Adams Whipple and the Bonds at the Harvard Observatory. British astronomers had photographed the moon, but the stars were many orders of magnitude smaller, and no one in England had succeeded in photographing them. The photograph that Mitchell carried of Mizar and Alcor (two stars in the Big Dipper) was the first of its kind seen in England. It was clear to all that photography of the stars would eventually make astronomers' tasks much easier. (Even today, close examination of photographs yields new information about planetoids and other wandering bodies.) Airy made his longing for the photograph so clear that Mitchell (who had planned to carry it through Europe, to show to anyone who was interested) impulsively gave it to him to keep at the Greenwich Observatory. Her generous impulse was the right one: the grand observatory at Greenwich became like a second home to her. She visited repeatedly, taking careful notes on the instruments and the setup, sleeping in a small round room "at the top of the little jutting out curved building."[17]

At Greenwich, Mitchell was at the geographical center of the world. The great transit circle in the observatory had crosshairs at the center of its field: when a star was centered on the crosshairs of the Greenwich transit circle, it was officially at the zero meridian. The transit circle was equipped with a telegraphlike electrical recorder. George Airy watched in his telescope sights as a star approached the meridian line, and the moment it did he pressed a button at the side of the transit circle to record the star's position. Airy decided where everything in heaven was, and since the Greenwich meridian was the basis for international timekeeping, he also decided *when* everything on earth was happening.

Although Mitchell grew very fond of Airy, she couldn't resist poking fun at him. She called him "the astronomical autocrat" as well as the "Bear of Blackheath," and she commented that he was "naturally a despot" whose inclination to bossiness was encouraged by his position. Sitting in his "throne," as she put it, he was at "the zero point of longitude of the world," commanding not only "the little knot of observers and computers around him," but also setting the clocks for every town hall and church tower in England and every ship's deck around the globe.[18]

Greenwich kept time for most of the world by the 1850s, although the compact that made Greenwich the international prime meridian would not be sealed until late in the century. But long before the international treaties were signed, burgeoning international trade made it urgently necessary for clocks to be synchronized. For ships, synchronizing time to a standard meridian was essential for figuring longitude and thereby navigating safely around the world. Sloppy timekeeping meant losing track of one's position and potentially crashing into shoals, islands, or even continental coastlines. Trains also needed careful timetables and accurate clocks—if railroad clocks were not carefully synchronized, trains might crash head-on along the carefully scheduled single rails that were beginning to crisscross Europe and North America.

Thinking about Airy's power to set the world's clocks and thus to keep England's trains running and her ships on course, Mitchell commented, "It is singular what a motive power Science is, the breath of a nation's progress." Everywhere she went, Mitchell compared Europe to America, trying to determine what European scientists did well and what American scientists might be able to do better. She loved her time at Greenwich, loved the fact that the Airys danced in the great room at the base of the old observatory and that there were little bedrooms tucked into the projections around its dome. Comfortable and inspired, Mitchell had found her ideal observatory. She looked out on the tall oaks of Greenwich Park, said to have been planted in Elizabeth's time; the Thames curved in a sharp oxbow directly west of the observatory; and a few miles beyond it she could see the perpetual cloud of smoke that hung over central London. The observatory is only four miles west of the Tower Bridge on the Thames, but at the time it was an almost rural oasis, where the skies were clear and dark enough for scientists to make fine observations.

In the autumn of 1857, Mitchell moved back and forth between London and Greenwich. The Airys helped her secure invitations to London's most fashionable salons—at a London rout hosted by the aristocratic Mr. Baden-Powell she was delighted to meet the physicist Neil Arnott and to catch a glimpse of Charles Babbage, whose "difference engine" was the mechanical precursor to today's digital computers. As a human computer, often wearied by the repetitive tasks of doing the complex calculations needed for every year's almanac, Mitchell must have longed for Babbage's engine to succeed.

It never did work—in fact, Babbage never really succeeded in building one—but the idea of the difference engine excited Mitchell nonetheless. Beyond London, Mitchell made forays to the Lake District, where she hiked energetically through the hills, and to Scotland, where she visited observatories at Edinburgh and Glasgow, always taking careful notes.

Thanks to the Airys, Mitchell also secured an invitation to visit William Whewell at Cambridge. Whewell was the master of Trinity College, where Isaac Newton had been a fellow, and an astronomer who had done extensive research on the tides, but he was actually more acclaimed as a historian and philosopher of science. Whewell had won prizes for his poetry as an undergraduate, and he continued to publish poems long after he became a fellow of the Royal Society and served as president of the British Association for the Advancement of Science. Whewell's love of language led him to coin words: Michael Faraday often consulted with him and asked for help in naming new things. Whewell came up with "ion," "anode," and "cathode," for example. Perhaps his best coinage was the word "scientist," which he first used in print in an 1834 essay praising Mary Somerville.

Because Whewell was so appreciative of Somerville, one would expect that he would have welcomed Mitchell as another distinguished woman scientist. Alas, the two disliked each other almost on sight. At their first meeting, he "was very severe upon Americans," going so far as to insult Ralph Waldo Emerson. Feeling that "his severity reached to discourtesy," Mitchell announced to him that she knew Emerson and "valued [her] acquaintance with him highly."[19] They got off on the wrong foot. Mitchell wrote, "Mrs. Airy said that Whewell and I *riled* each other." Mitchell found Whewell unbearably stuffy and his love of academic regalia simply ridiculous. He, on the other hand, was nonplussed by her, shocked by her vulgar taste in poetry—he couldn't believe she preferred Elizabeth Barrett Browning's passionate political verses to Felicia Hemans's delicate sentimentality—and outraged by her lack of reverence for Cambridge.

Cambridge University made Maria Mitchell feel merry and mischievous. Perhaps she was overwhelmed by its exclusivity; certainly she found the manners of Cambridge dons a bit offensive, and she thought Whewell's scarlet robes were just plain silly. Mitchell admired the fact that Cambridge was the first university to set up organized programs in the physical sciences and the fact that Newton's legacy remained strong—but the stultifying atmosphere of aristocratic male privilege remained strong as well. Mitchell had a good talk with John Couch Adams, the calculator of Neptune's orbit, and took away hilarious anecdotes of Whewell's rudeness, but she definitely preferred Cambridge, Massachusetts, to Cambridge, England.

Excluded from a university chapel more than fifty years later, Virginia Woolf seethed outside and bitterly described the scene: "[I]t was amusing enough to watch the congregation assembling, coming in and going out again, busying themselves at the door of the chapel like bees at the mouth of a hive. Many were in cap and gown; some had a tuft of fur on their shoulders; others were wheeled in bath-chairs; others, though not past middle age, seemed creased and crushed into shapes so singular that one was reminded of those giant crabs and crawfish who heave with difficulty across the sand of an aquarium."[20] Mitchell was a guest of the master of Trinity College, so she was not excluded, but her sense of the absurdity of the scene is parallel to Woolf's:

> Sitting at the window of the hotel, [a visitor] will see scholars, the fellows, the masters of arts, and the masters of colleges passing along the streets in their different gowns. Very unbecoming gowns they are, in all cases; and much as the wearers must be accustomed to them, they seem to step awkwardly, and to have an ungraceful feminine touch in their motions. Everything you see speaks of the olden time.... The costumes of Cambridge and Oxford are very amusing, and show, more than anything I have seen, the old-fogyism of English ways. Dr. Whewell wore, on this occasion, a long gown reaching nearly to his feet, of rich scarlet and adorned with flowing ribands. The ribands did not match the robe, but were more of a crimson. I wondered that a strong-minded man like Dr. Whewell could tolerate such trappings for a moment, but it is said that he is rather proud of them, and loves all the etiquette of the olden time, as also, it is said, does the Queen. In these robes, Dr. Whewell escorted me to church—and of course we were a great sight![21]

Mitchell had always been self-conscious about her appearance. Her trip to England was particularly hard on her sartorially; at one point she was castigated by strangers in a theater for wearing a bonnet rather than a hood (apparently bonnets were permitted in the pit only), and everywhere she went her old-fashioned Quaker garb marked her as an exotic. But she was stubbornly determined to stick to her plain style, never deviating from the muted Quaker palette of gray, brown, and black. The British were much more colorful—especially the undergraduates at Cambridge, fluttering in blue silk robes around the crimson-ribboned scarlet robes of their masters. To walk through crowds of such men in her own dull garb must have made her acutely aware of her own exclusion.

In her journals and letters, Mitchell played down her sense of being an outsider, and although she poked fun at Whewell, she avoided expressing anger toward him or his stodgy institution. Instead, she summed up Whewell's book on the plurality of worlds with humor: "The planets were created for this world; this world for man; man for England; England for Cambridge; and Cambridge for Dr. Whewell!"[22]

Beneath Mitchell's laughter, there was a great deal of frustration. She was a provincial scholar from a developing country. "Nothing is more provoking than the ignorance of the English about Americans," she declared, noting, "In the opinion of a Cambridge man, to be master of Trinity is to be master of the world!"[23] Her sense of Whewell's central position in the scientific universe was at least as strong as his own. But although the old-world snobbery of academic England offended Mitchell's Yankee sensibilities, it also made her feel optimistic for American science. When she visited the Cambridge observatory she noted merely that "Professor Challis, the director, is exceedingly short, thick-headed (in appearance), and like many of the English, thick-tongued."[24] When she met a scholar she liked, however, she praised his "American" qualities. After meeting Dr. Toynbee, who introduced himself by mentioning that he admired Emerson, she remarked that he was "full of enthusiasm and progress—like an American. He really seemed to me all alive, and is either a genius or crazy—the shade between is so delicate that I can't always tell to which a person belongs!"[25] Most British scientists seemed dull by comparison. Later Mitchell explained the importance of studying the British national mind-set:

> There is a phenomenon well-known to astronomical observers as "personal equation." No two persons receive an impression and make it

known in the same time. Thus, if one sees a star, and calls out that he sees it, the interval of time which elapses between the sight and the call, the seeing and the speaking, is different for any two persons. We call this difference "personal equation."

There seems to be a "national equation." We do not expect that even the little popular scientific work which we take up written in French shall reach conclusions by the same process as those by which the little German book will reach the same. If we would understand, thus, the science of the period, we must know the national soils in which science has taken root.

A singular illustration of national difference was seen in the case of the discovery of the planet Neptune. Two leading men, one in England, one in France, sitting in their studies, proved by careful mathematical investigations that there must be a planet away out beyond what were considered the bounds of the solar system. The Englishman worked out his problem first, but pondered long, thought much, and consulted with others before he published it. The Frenchman finished his computation, put his pencil down, and announced the result in the next day's papers. But a third, and he was an American, said, "True, you have each declared a planet to exist, and a planet has been found, but you did not agree in your calculations, and the planet which has been found is not the planet announced by either."[26]

The men Mitchell referred to were her almanac colleague Sears Cook Walker, the English astronomer John Couch Adams, and the Frenchman Urbain Leverrier. She met both European astronomers in the fall of 1857, each on his home turf, and had pleasant exchanges with both. But as she spoke to them, she was also developing her ideal for American astronomy. As her remarks suggest, she thought of the British as slow, overly theoretical, and somewhat hidebound. The French seemed too impulsive. American astronomers, Mitchell thought, might be the best of the lot, if they could hang on to their enthusiasm and combine it with hardheaded pragmatism and unflinching accuracy. She was fairly certain that questions of social class limited British astronomers. She respected the fellowship system at the universities, which allowed admission to a few working-class boys, but she was appalled by the critical role that the aristocracy played in British astronomy. "The amateur astronomers of England are numerous, but they are not like those of America," she remarked.

In America, a poor schoolmaster who has some bright boys who ask questions, buys a glass and becomes a star-gazer, without time and almost without instruments; or a watchmaker must know the time, and therefore watches the stars as timekeepers. In almost all cases, they are hardworking men. In England, it is quite otherwise. A wealthy gentleman buys a telescope, as he would buy a library, as an ornament to his house...If he is a man of philosophical mind he soon becomes an astronomer, or if a benevolent man, he perceives that some friend in more limited circumstances might use it well, and he offers the telescope to him, or if an ostentatious man he hires some young astronomer of talent, who comes to his observatory and makes a name for him. Then the queen confers the honor of knighthood, not upon the young man, but upon the owner of the telescope.[27]

As Mitchell continued on her journey, her frustration and outrage were balanced by a growing sense of American potential. Before her travels, she had thought of her situation as seriously disadvantaged, but once she became acquainted with leading European astronomers, she was able to see that an American astronomer who knew the "national soils" in which the science of other lands had taken root was much less limited than a European scientist bound by caste and national history. The opportunities for American astronomy, Mitchell was coming to believe, were endless.

Late 1857 was a very exciting time for Mitchell—the world was now far more open to her than she had expected. When she met the great Russian astronomer Wilhelm Struve, who was also visiting England, she mentioned that she carried a letter of introduction to him and that she had planned to make her way to St. Petersburg to visit him at the Pulkova Observatory. She was pleasantly surprised when he responded that there was no need for a letter—he knew her by reputation.

But while Mitchell was accumulating this positive experience, her traveling companion, Prudence Swift, was having a terrible time. That fall, her father, an American businessman, lost his fortune in a financial crash that impoverished many Americans, and Prudie was suddenly too poor to travel. Mitchell arranged for her to sail home in early November and probably felt great relief. Prudie might have been relatively pleasant, but the two women had few common interests; Swift was fifteen years younger than Mitchell and much more interested in suitors than in stars. Once Prudence was safely packed off to New York, Mitchell was free to devote all her time to her enlarging her circle of astronomical acquaintances.

Years later, Maria Mitchell's sister Phebe remembered that when the young Mitchell children were asked, "Who is the greatest man that ever lived?" they all had the same instant response: "Herschel!"(presumably meaning William Herschel, who had discovered Uranus in 1781). There is an interesting ambiguity to Phebe's use of the surname alone, because there were three Herschels who interested the Mitchells. They might also have been referring to William's son, John Herschel, who wrote some of the most significant astronomical treatises of the nineteenth century and whom Mrs. Airy described as "the acknowledged head of astronomy" by the mid-nineteenth century.[28] Finally, there was Caroline Herschel, William's sister and John's aunt.

Maria Mitchell was probably much more curious about Caroline, who died in 1848 shortly after she discovered her own comet, than she was about John, who was living with his family at Collingwood in 1857. Every detail about Caroline intrigued her—Mitchell learned, early in her visit, that Caroline was famous for wearing fourteen pairs of stockings at once when she observed late at night. Mitchell ruefully acknowledged the humor in this detail, but she also stoutly commented that "when she wrapped herself in innumerable wrappings and took care of the body that the mind might do its duty, she gave a lesson which every girl ought to follow."[29] But Mitchell's curiosity about Caroline Herschel went far beyond her stockings. Caroline Herschel and Mary Somerville were the only two women who had been formally recognized by the Royal Society; they were Mitchell's two most important role models as women. Richarda Airy had put Mitchell in touch with the Herschels, and Maria was thrilled to have secured an invitation to visit them at Collingwood. Caroline and William, the two members of the family about whom she was most curious, were both long departed; when William died in 1822, Caroline moved to Germany, where she died in 1848, two years short of her hundredth birthday. By the time Mitchell visited the Herschels in 1857, Caroline had been gone from England for thirty-five years. Even so, it was Mitchell's first chance to be close to the relics of a world-renowned woman astronomer. The Herschels welcomed Mitchell, and she enjoyed their company immensely. John Herschel was sixty-six and no longer as scientifically active as he had been early in his career, but he was still very well informed, and he continued to revise and publish his writings. Mitchell was pleasantly surprised by his manners: she had expected another tyrant, like Airy (whom she adored, despite his despotic tendencies) or Whewell (who offended her). Instead, she found John to be "womanly in nature," "a better listener than any man I have met in England." She was delighted by the way he joined in all the "chit-chat" of the domestic circle.[30] The days and nights spent at Collingwood were pleasant.

As Mitchell was saying farewell, Sir John bustled up to her and handed her a small sheet of paper torn from the observatory records. This little scrap of Caroline Herschel's handwriting became one of Mitchell's most treasured possessions. Each sentence started with the initials *W.H.* (for William Herschel): "W.H. says ... ," "W.H. thinks ... ," "W.H. has put it in both zones ... ," and so on. The initial *C*, for Caroline, is nowhere. When Mitchell pasted it into her scrapbook, she labeled it, "Notes made by Caroline Herschel for her brother, Sir Wm. Herschel."

As Mitchell fingered the paper, she thought about Caroline's astronomical career. Although Caroline had loved to observe on her own, the scrap was from one of the many nights when she had sat in the dark at her brother's elbow, carefully recording what he saw. Later at night, after William had gone to sleep, Caroline would take her own small comet sweeper down from its shelf and look around for herself; in the morning she would be up early to begin the necessary calculations.

The nineteenth-century historian of science Agnes Mary Clerke describes Caroline's routine this way: "Busy days succeeded watchful nights. From the materials collected at the telescope, she formed properly arranged catalogues, calculating, in all, the places of 2,500 nebulae. She brought the whole of Flamsteed's British Catalogue—then the vade mecum of astronomers—into zones of one degree wide, for the purposes of William's methodical examination; copied out his papers for the Royal Society; kept the observing books straight, and documents in order. Then, in the long summer months, when 'there was nothing but grinding and polishing to be seen,' she took her share of that, too."[31] Only when her brother was absent was she free to search for comets. Clerke comments, "Considering that she swept the heavens only as an interlude to her regular duties, never for an hour forsaking her place beside the great telescope in the garden, her aptitude for that fascinating pursuit must be rated very high."[32] Clerke concludes, "Caroline Herschel was the first woman to discover a comet; and her remarkable success in what [Fanny] Burney called her 'eccentric vocation' procured for her an European reputation. But the homage which she received did not disturb her sense of subordination."[33]

Caroline Herschel's extraordinary "sense of subordination" was troubling to Maria Mitchell. It seemed "ungenerous to blame at all where we admire so much," but she could not wholeheartedly endorse Caroline's unflagging self-denial.[34] Although she reflected on Herschel for years, she hesitated to publish her opinions; it was only after her death, in 1889, that Mitchell's thoughtful discussion of Caroline Herschel was published in the *Century*.

Mitchell admired Herschel's intelligence and talent, but she came to believe that Herschel had shirked the responsibility that her gifts imposed upon her. As Mitchell put it,

> If you had asked Caroline Herschel after ten years of labor what good had come of it, she probably would have answered, with the extreme simplicity of her nature, that she had relieved her brother of a good deal of wearisome labor, and perhaps kept up his vigor and prolonged his life. Probably it never occurred to her to be other than the patient and self-sacrificing assistant to a truly great man.
>
> The woman who has peculiar gifts has a definite line marked out for her, and the call from God to do his work in the field of scientific investigation may be as imperative as that which calls the missionary into the moral field or the mother into the family: as missionary, or as scientist, as sister or as mother, no woman has the right to lose her individuality. To discuss the question whether women have the capacity for original investigation in science is simply idle until equal opportunity is given them. We cannot overrate the consequences of such lives, whether it be Mrs. Somerville translating LaPlace, Harriet Hosmer modeling her statues, Mrs. Browning writing her poems or Caroline Herschel spending nights under the open canopy; in all it is devotion to idea, the loyalty to duty which reaches to all ages.[35]

Caroline Herschel made a mistake, Mitchell thought, when she put her devotion to her brother above her devotion to her own ideas: she lost her individuality. In anyone's life this would be a great loss, but when Caroline Herschel made this choice, others lost out as well— who knows what she might have discovered, or what help she might have offered to the scientifically inclined women who would follow her? "When she worked in the little observatory at Slough," Mitchell wrote, "she not only worked in every observatory of the world, but she reached to every school for girls. If what Caroline Herschel did is a lesson and a stimulus to all women, what she did not do is a warning. When Caroline Herschel so devoted herself to her brother that on his death her own self died and her life became comparatively useless, she did, all unconsciously, a wrong, and she made the great mistake of her life." Yet Mitchell did acknowledge, "The fault was only in part her fault." Circumstances were against Herschel; her family and her culture valued and celebrated her self-sacrifice to such a degree that it would have been difficult for her to relent toward herself. Even so, Mitchell's plangent question remains: "Has any being a right not to be?"[36] Herschel's example was, finally, a negative one: she was a reverse role model, a woman whose life was more cautionary tale than inspiration.

Mitchell nonetheless treasured the scrap of Herschel's handwriting and valued her acquaintance with Caroline's nephew and his family. John Herschel was an important astronomer and his contributions to the development of photography had significance far beyond astronomy, but Mitchell probably admired Herschel most for his writing: his *Preliminary Discourse* (1831) was as highly regarded in America as it was in Europe. Interestingly, it was one of Ralph Waldo Emerson's favorite books because it articulated an approach to nature that blended Romanticism with natural philosophy.[37] John Herschel was a formidable historian and philosopher of science, and his ideas on the connections among philosophy, literature, and science remained important for Mitchell throughout her career. William Whewell was a disciple of John Herschel, but Whewell led the move away from natural philosophy and toward a discipline based solely on inductive methodology that eschewed the other (perhaps more philosophical) practice of deduction. Mitchell recognized that inductive science was growing away from the humanities and that the great advances were being made by the most focused scholars, but her own sensibilities had been shaped in an earlier time: she was searching for the successors to the Herschels. Nonetheless, she admired the new scientists all the same. On leaving Collingwood, she wrote to her father, "I am hoping to get to Paris next week, about the 23rd. I have had just what I wanted in England, as to society."[38]

During her three months in England, Mitchell had met dozens of prominent scientists. Sometimes she only managed a glimpse, as in the case of Charles Babbage, the builder of the computing engine. At other times, she spent more time defending America than discussing science, as in the case of Whewell at Cambridge. But she had, in the end, met Adams and Struve, Lassell and Babbage, and had established real friendships with Smyth, Airy, and Herschel. The social connections she had made were invaluable. Her letters do not mention meeting Joule, von Hoffmann, Maxwell, Lyell, Darwin, or Thomson, Lord Kelvin. These cutting-edge scientists of the 1850s may have been a bit too obscure at the time to merit mention, or perhaps Mitchell was yet uninformed about the new and profound developments in electromagnetism, thermodynamics, and natural selection.

As she set out for Paris, Mitchell was beginning to feel that she had her finger on the scientific pulse. She would never be a fat spider at the center of the web like Whewell, but she might become a woman scientist different from Caroline Herschel—one who could chart out her own course of research. She was also reassured by the family life she had seen. In the early 1850s, she had worried that leaving Nantucket to become a professional astronomer might turn her into a public figure with no connection to a home circle, but she was finding that the observatories of the world were new homes for her—strangely furnished with telescopes and transit circles, and strung with telegraph wires, but merry and pleasant nonetheless. The home circles of the Airys, the Herschels, and the Smyths were all warm, comfortable, loving, and welcoming. Mitchell was beginning to imagine a multidimensional life for herself as a professional astronomer, at home in her own observatory.

CHAPTER 7

The Yankee Corinnes

When Maria Mitchell left England, she left the center of the astronomical world. It had been fascinating to study the English scientific establishment while it was in the process of establishing itself, and most of the English astronomers had been very kind to her, but her experience in England troubled her in many ways. She had always been aware of her status as a *woman* astronomer, and to her chagrin she ultimately found that her colleagues tended to speak to her as a woman rather than as an astronomer. In England, Mitchell learned, few girls or women attended organized schools—their education was private and relatively haphazard, and intellectual women were considered decidedly strange. After a few months of trying to negotiate the surprise and curiosity of British scientists, Mitchell was filled with questions about her purpose. She set her course for the Continent.

Mitchell was now living outside Nantucket for the first time in her life. Nantucket was full of educated women, and the United States in general was much more invested in women's education than England was. Though her hometown was exceptionally encouraging to women's intellectual curiosity, there were schools and seminaries for girls all over the United States, most staffed by women teachers.

Mitchell conscientiously visited astronomers everywhere she went, but she was equally interested in the women of genius who peopled the Paris, Rome, and Florence of her imagination. The writers Madame de Staël and George Sand fascinated her; she admired the painter Rosa Bonheur; and she couldn't stop wondering about the Italian experiences of Margaret Fuller and Elizabeth Barrett Browning. Most of all, she looked forward to meeting Mary Somerville, the acclaimed Scottish scientist and mathematician who lived in Florence.

Another reason Mitchell's attention was shifting away from astronomy was that her astronomical connections on the Continent were much less interesting than they had been in England. She met the astronomer Urbain Leverrier a few times at his salons, and he took her on a cursory tour of the Imperial Observatory, which he directed. Leverrier was at odds with the British astronomical community because of the dispute between Adams and himself over the discovery of Neptune, and both Mitchell and Leverrier were ill at ease. Communication was difficult; Mitchell commented, "The Le Verriers speak English about as well as I do French, and we had a very awkward time of it."[1] Even so, Mitchell was welcomed more generously than Margaret Fuller had been ten years before, when the guards had refused to allow her into the Sorbonne to hear Leverrier lecture.[2] Yet although Leverrier was reasonably polite to Mitchell, he didn't invite her into the observatory's dome. She thought, "It was evident that he didn't much expect me to understand an observatory."[3] Privately she returned his low opinion, writing, "If what he showed me is not surpassed in the other rooms, I don't think much of their instruments."[4] Perhaps observatories were all beginning to seem alike to her. She complained in her journal, "All the observatories of Europe seem to have been built as temples to Urania and not as working chambers of science."[5] Disappointingly, many of the astronomers she visited seemed to be operating on an older model of astronomy in which it was more an art than a science, and observatories were more like churches than laboratories.

Paris was blessedly quiet after the urban intensity of London. Mitchell was not surprised—she had read Harriet Beecher Stowe's accounts of Paris, so she knew the Parisian air would be, in Stowe's words, "so different from the fog and smoke of London. There is more oxygen in the atmosphere. A pall is lifted."[6] Though it was bitterly cold, and some days were filled with hard, driving December rain, Mitchell agreed with all of the guidebooks that Paris was splendid. She was a dogged tourist, sightseeing "from ten in the morning till ten at night,"[7] and she loved the wide boulevards and the sense of spaciousness at the city's center, commenting approvingly "in Paris you have room to look at things."[8] In contrast to London, Mitchell wrote, "The city is as quiet as Nantucket."[9] Mitchell also felt somewhat frightened by Paris. She was relieved that Prudence Swift had returned to America because Paris seemed dangerous to a young girl's morals. Harriet Beecher Stowe had warned of this problem when describing the view from the Arc de Triomphe: "Paris was beneath us, from the Louvre to the Bois de Boulogne, with its gardens and moving myriads; its sports, and games, and lighthearted mirth—a vast vanity Fair, blazing in the sunlight. A deep and strangely-blended impression of sadness and gayety sunk into our hearts as we gazed. All is vivacity, gracefulness, and sparkle, to the eye; but ah, what fires are smouldering below! Are not these vines rooted in the lava and ashes of the volcano side?"[10] The year that Mitchell visited, Charles Baudelaire published *Les Fleurs du Mal* (*The Flowers of Evil*), a book of poems that were rooted in precisely the smoldering fires that frightened New England women such as Stowe and Mitchell.

Perhaps because of her anxiety about the atmosphere of sexual freedom, or perhaps simply because her French was weak, Mitchell seems not to have tried to meet George Sand or Rosa Bonheur while she was in Paris. It is reasonable to imagine that Sand would have been a figure of interest to Mitchell; both Margaret Fuller and Elizabeth Barrett Browning had sought out the novelist when they visited Paris. Perhaps Mitchell was put off by Sand's publicly acknowledged love affairs, but she would probably have taken Sand's preference for male attire in stride. Rosa Bonheur was also wandering around Paris in men's clothing at the time, with a certificate of permission to dress as a man granted her by the local police. As Maria Mitchell visited the galleries of the Louvre, walked along the banks of the Seine, or strolled through the grand courtyard of the Institut de France, she was often surprised by the low-cut dresses of Parisian women, and she was always aware that any of the gentlemen quietly painting or casually passing by might be a woman of genius in disguise. The idea of women in trousers still shocked nearly everyone in Paris because tailoring revealed the shape of women's legs, but by Nantucket standards, half-bare bosoms were at least as shocking. No matter where she turned in Paris, Mitchell was forced to think about women's bodies.

Mitchell came from a family that was extremely modest. According to the Mitchell biographer Helen Wright, one of Mitchell's sisters "in her last illness, confessed very devoutly her thankfulness that her husband had never seen her naked." In the galleries of the Louvre, there were images of naked women everywhere. Mitchell left no written comments about them, but Stowe, whose travel book Mitchell praised, overcame her New England attitudes to advise travelers to enjoy the Rubens: "his women shall be as fat as he pleases, and you shall like him nonetheless!" Stowe saved her strongest praise and most careful thought for Venus de Milo. Comparing the statue to Milton's Eve, she commented that "there is a majesty and grace in the head and face, a union of loveliness with intellectual and moral strength, beyond anything which I have ever seen."[11] As Mitchell stood alone in the Louvre looking at the nude torso, her thoughts may well have turned toward the planet Venus, her responsibility in the *Nautical Almanac*. In Europe, the women of ancient mythology—Venus, Athena, and Urania in particular—were becoming real to her in new ways.

On top of the classical ideals, there was a new, nineteenth-century ideal: Corinne. Corinne was the heroine of Madame de Staël's 1807 novel—a great beauty who won fame and honor for her genius at improvising poems and declaiming her verses to the crowds who thronged around her on the hills of Rome. Knowing that Margaret Fuller had been pleased and proud when Ralph Waldo Emerson and James Freeman Clarke (Fuller's eventual biographer) called her a "New England Corinne," Mitchell was eager to see the city of Rome, where both the fictional Corinne and the real Margaret Fuller had thrived.

Mitchell had never thought of herself as a beauty. Still, de Staël's novel shaped her travel to Rome as it shaped the journeys of most Anglo-American women in the nineteenth century. Ralph Waldo Emerson looked to de Staël for his understandings of science, Margaret Fuller embraced the identity of a Yankee Corinne, and Elizabeth Barrett Browning made a heroic effort to emulate the fictional Corinne as she shaped her own life as an English woman poet in Italy. Even if Mitchell hesitated to think of herself as a Corinne, two of the women whose work she admired the most, Fuller and Browning, both imagined themselves that way.

De Staël was a Frenchwoman who had been exiled during the

revolution. Her novel tells the story of the artistic coming of age of a talented and beautiful, half English, half Italian woman who is exiled to Italy. It is an unhappy tale—the English lord who loves her in Italy eventually abandons her for her purer, less imaginative sister in England. But generations of eager intellectual women ignored the tragic ending, finding inspiration in Corinne's early poetic success. A few such readers traveled to Italy, at least in part because they wanted to emulate de Staël's heroine; Margaret Fuller, Elizabeth Barrett Browning, and Julia Ward Howe were three women who had made the journey. Mitchell may never have met Fuller, and she certainly never met Barrett Browning, but when she went to Italy, she was following in their footsteps and trying to imagine herself among anglophone women of genius.

A remote colony of the Austrian empire, Rome was at this time a small, mostly ruined city dominated by a Catholic Church establishment that actively excluded women. Rome had history, of course, and art and antiquities galore, but it did not have much of a history of opportunities for women before de Staël set her novel there. Perhaps one reason that the story of Corinne was so powerful was that it portrayed the woman of genius as an exile. In the nineteenth century, to be female was to be only partially a national citizen. Women did not have the vote anywhere in the world, and did not necessarily carry independent passports; their citizenship status was often wholly dependent on that of their male relations. Since Rome was far removed from Austria, which had nominal control over the region, the city was easy to reimagine as a concrete landscape of cultural history, somehow separate from nationalism.

But Italian nationalism was actually at fever pitch during the years of the Risorgimento, from the time of Browning's and Fuller's first journeys to Italy through to Mitchell's trip. Both Browning and Fuller supported the Italian revolutionaries wholeheartedly and worked for the Risorgimento, Browning by writing a series of poems about the revolution in Florence, Fuller by writing essays for the *New York Tribune* and by working as a nurse to wounded revolutionaries. Italians were reinventing themselves as national citizens.

The politics of the Risorgimento and the story of Corinne both made Italy seem like a place of great opportunity for women. The immensely powerful romantic travelogue poems of George Gordon,

Lord Byron, also linked women to Italy, and every English speaker who visited Italy seemed to carry a well-thumbed copy of Byron's narrative poem "Childe Harold's Pilgrimage." Back in 1854, Mitchell had met James Freeman Clarke when he gave a lecture on Italy at the Atheneum. Mitchell had written at the time, "In the evening, I listened to him lecture. It was 'a month in Italy,' and contained nothing that we had not heard before. He concluded with some verses of Byron which he read beautifully. All travelers to Europe seem to make Byron their *vade mecum* on the journey, and most prefer him"[12] to any guidebook. When she went to Europe, she made sure to bring her own copy of Byron.

Most Anglo-American travelers at the time saw continental Europe through Byron's eyes, and his vision of Rome as a ruined matriarch seems to have added fuel to the feminist imaginations that saw Italy as a place of great possibility. Though the descriptions of ruins in "Childe Harold's Pilgrimage" are often compelling, Byron makes an odd travel guide, especially for a woman, since much of his poem is about romantic masculinity. Nonetheless, there is something quite feminine about Byron's Rome. He addresses Rome this way:

> Oh Rome! my country! city of the soul!
> The orphans of the heart must turn to thee,
> Lone mother of dead empires! [13]
>
> . . .
>
> The Niobe of Nations! There she stands
> Childless and crownless, in her voiceless woe.[14]

In Greek mythology, Niobe is the figure of the mourning mother. The gods killed her fourteen beautiful children to punish her for being proud of them. Byron's description of Rome as the Niobe of nations alludes to the city as an ancestor of many empires—but it also makes Rome startlingly female. For the Yankee Corinne, Margaret Fuller, and for the British one, Elizabeth Barrett Browning, the heady combination of Byron and de Staël made Italy, and particularly Rome, the site of a powerful fantasy of female intellectual fulfillment.

Foreign travel at that time was onerous and expensive. Few could afford the process of hiring horses and carriages and organizing travel around the Continent, and even fewer single women could manage it. Still, motivated by their powerful imaginations, Elizabeth Barrett Browning and Margaret Fuller were determined to get there. Barrett Browning eloped with Robert Browning, and they traveled together; a few years later, Fuller made the journey on her own, though she traveled with a married couple, Rebecca and Marcus Spring.

Fuller, having spent her childhood studying the classics with her father, knew a great deal of Latin prose and poetry by heart, in addition to the works of Byron and de Staël. Mitchell was a decent classicist herself; she was one of the only American astronomers who regularly used astronomical treatises written in Latin. To gain such familiarity with the language she must have spent some time reading Caesar, Virgil, and Cicero. Her ideas about Italy were probably also shaped by many other nineteenth-century writers who had written about the place—Goethe, Dickens, Trollope, and, in the United States, Washington Irving and Margaret Fuller, not to mention the many travel writers who elaborated on the others' descriptions. According to the scholar Richard Brodhead, American tourism in Europe before the Civil War was relatively limited but gradually increasing. Brodhead reports that "one possibly reliable count lists two hundred to three hundred Americans passing through Rome in 1835, but a thousand by 1840."[15] By 1867, shortly after the Civil War, numbers had swelled, exceeding a thousand a week at times. So in 1858, as Mitchell traveled from Paris to Rome, she was on an exclusive but well-traveled road.

On January 9, 1858, Mitchell visited Nathaniel and Sophia Hawthorne, who were also staying in Paris, to ask whether she could join them on the difficult journey from Paris to Rome. Mitchell knew that both Nathaniel and Sophia had been very close friends of Margaret Fuller's, and she was aware of rumors that there had been a painful rupture between the Hawthornes and Fuller—rumors that seemed to be supported by the harsh portrait of a Fuller-type heroine in Hawthorne's 1852 novel *The Blithedale Romance*.

Nathaniel Hawthorne laughingly commented in his journal, "This morning, Miss Mitchell, the celebrated astronomical lady of Nantucket, called. She had brought a letter of introduction to me, while Consul, and her business now was to see if we could take her as one of our party to Rome, whither she likewise is bound. We readily consented; for she seems to be a simple, strong, healthy-humored woman, who will not fling herself as a burden on our shoulders; and my only wonder is, that a person evidently so able to take care of herself should care about having an escort."[16]

Mitchell did prove more competent than the Hawthornes in the logistics of travel, using her rudimentary French to help them communicate with innkeepers and maidservants and entertaining the children by feeding them gingerbread, telling them stories, and even teaching them a little about the stars. On the voyage toward Rome, Nathaniel Hawthorne recorded, "The evening was beautiful, with a bright young moonlight, not yet sufficiently powerful to overwhelm the stars; and as we walked the deck, Miss Mitchell showed the children the stars and constellations, and told their names. Julian made a slight mistake as to one of these, pointing it out to me as 'O'Brian's Belt.'"[17] Mitchell's travels with the Hawthornes became the basis for a solid family friendship. When they arrived in Rome, she rented her own rooms but continued to see them often, dining in their apartment and doing a great deal of touring with them. She and Sophia Hawthorne developed a warm friendship, and the children loved her. Many years later, Rose Hawthorne recalled that Mitchell "smiled blissfully in Rome, as if really visiting a constellation; flashing her eyes with silent laughter, and curling her soft, full, splendid lips with fascinating expressions of satisfaction. I loved her for this."[18]

But the Hawthornes were strange companions for Mitchell as she explored Corinne's Italy. Thoughts of Margaret Fuller were hard for Mitchell to keep at bay, but Nathaniel and Sophia probably wanted to avoid the subject. When Mitchell asked about Fuller, Sophie told her "that Mr. Hawthorne admits that in the character of Zenobia" in *The Blithedale Romance* "he felt Margaret Fuller's presence." Mitchell got the impression from Sophia that Fuller's presence was unwelcome. "She says no one loved Margaret Fuller," Mitchell wrote, "... that Mr. Hawthorne felt her need of women's refinements and was disgusted."[19] Hawthorne himself was more outspoken. Writing in his journal during the ensuing winter in Rome, he remarked that Fuller "was a woman anxious to try all things, and fill up her experience in all directions; she had a strong and coarse nature, too.... Margaret has not left, in the hearts and minds of those who knew her, any deep witness for her integrity and purity. She was a great humbug."[20] Hawthorne thought that Fuller's liaison with Ossoli proved that she had "a defective and evil nature," and he thought that her resolve to "make herself the greatest, wisest, best woman of the age" was "an awful joke," because she "proved herself a woman after all, and fell as the weakest of her sisters might."[21] The Hawthornes' prudish disapproval of Fuller probably made an impression on Mitchell, but she gave no sign of sharing it and might not actually have known the depths of Nathaniel's revulsion. Years later, when Hawthorne's Italian journals were published (with their frequent, polite references to Miss Mitchell and their strong invective against Margaret Fuller), Mitchell wrote, "I have been surprised to see that he made some severe personal remarks in his journal, for in the three months I knew him, I never heard an unkind word; he was always courteous, gentle, and retiring."[22] Presumably, all three travelers kept their thoughts about the departed Margaret Fuller to themselves. For Mitchell, Fuller remained an abstract and distant figure, and the questions that Fuller had raised in *Woman in the Nineteenth Century* remained central to her thoughts.

Although Margaret Fuller had long been an object of Mitchell's fascination, Mitchell might not have been aware that her favorite poem, Elizabeth Barrett Browning's epic *Aurora Leigh,* was based in part on Margaret Fuller, whom Barrett Browning had befriended in the year before her death. The poet had seen Fuller as a tragic heroine. If *Aurora Leigh* is one response to Fuller's sojourn in Italy, Nathaniel Hawthorne's last novel, *The Marble Faun,* is surely another. *The Marble Faun* centers on the mysteriously independent American women artists whom Hawthorne met in Rome that winter, but he returned again and again to the question that so captivated Mitchell and Browning and that Fuller embodied outright: how can a woman be independent and intelligent and still remain a woman? It is unproductive to read any particular fictional character as a straightforward depiction of Fuller; Aurora Leigh is certainly not a Fuller manqué, and Miriam, the dark lady of *The Marble Faun,* is also not a simple portrait of her. But in both cases, Sophia Hawthorne's response to Mitchell's question about Zenobia seems relevant: "I asked her if Zenobia was intended for Margaret Fuller. And she said 'No,' but Mr. Hawthorne admitted that Margaret Fuller seemed to be around him when he wrote it."[23]

Fuller's ghost must have hovered around all three of them—Nathaniel, Sophia, and Mitchell—as they met in the Barberini Palace and looked together at Guido's portrait of Beatrice Cenci. In Rome in 1857, copies of Beatrice Cenci's portrait were as ubiquitous as copies of the *Mona Lisa* are in Paris today. The image was everywhere and would soon take a central role in Hawthorne's novel. There are many contradictory stories about Beatrice Cenci. Guido's portrait of her shows a young girl who is on her way to be executed after colluding in the murder of her father, who had raped her. It is as hard to explain the mid-nineteenth-century obsession with Cenci as it is to explain the current interest in the *Mona Lisa,* but it seems clear that for the Hawthornes, Beatrice Cenci offered a chance to consider women's sexual innocence and experience. The question of sexual purity was profoundly interesting to the Hawthornes and perhaps to Mitchell as well, though none of them would have discussed it. They simply gazed at the painting, and Nathaniel and Sophia later recorded their private reflections in their journals. When Nathaniel mined his journals to describe the portrait in *The Marble Faun,* he had two women look at the portrait together without the presence of a man. In some loose way, these two fictional women, Hilda and Miriam, can be connected to his wife, Sophia, and to the formidably independent Maria Mitchell.

Strangely, in Hawthorne's novelistic imagination, the ghost of Margaret Fuller seems to have hovered around Maria Mitchell. In *The Marble Faun,* he linked the character Miriam to his traveling companion Maria, not only by giving her a similar name but also by referring obliquely to Mitchell in his description of her, when he compared Miriam's eye to "the woman's eye that has discovered some new star." Ironically, his brief mention of the astronomer depicts her sewing, the single activity that Mitchell loathed most. In his description of the artist Miriam mending her glove, he wrote, "A needle is familiar to the fingers of them all. A queen, no doubt, plies it on occasion; the woman poet can use it as adroitly as her pen; the woman's eye, that has discovered some new star, turns from its glory to send the polished little instrument gleaming along the hem of her kerchief, or to darn a casual fray in her dress."[24] It would have horrified Mitchell to be connected with Miriam's dubious morality, sexual and otherwise, but it would have infuriated her to be depicted with a needle in her hand. Mitchell's tart response to Hawthorne's musings about women's needlework came in her 1860 profile of Mary Somerville in the *Atlantic Monthly,* where she wrote that "the needle, which had been the fetter of so many women, became, . . . [in Somerville's] hand, magnetic, and pointed her to her destiny."[25] Counter to Hawthorne's assertion that a "needle is familiar to the fingers of them all," and that sewing connects all women to domesticity, Mitchell argued that some women magnetize their needles into compass needles and use them to chart original courses for themselves.

But the ironies and complexities of Mitchell's relationship to Nathaniel Hawthorne go beyond their divergent attitudes toward needlework. That winter in Rome, the larger question for Mitchell and the Hawthornes concerned their attitudes toward women, and particularly women of genius. Mitchell may have been "a simple, strong, healthy-humored woman" but she was also a woman of intellect, and as such she was something of a threat to Hawthorne's troubled imagination. They seem to have gotten along well enough —they certainly never argued—but Mitchell once wrote, "Generally he sat by an open fire, with his feet thrust into the coals and a volume of Thackeray on his knees. . . . I sometimes suspected that the volume of Thackeray was kept as a foil, that he might not be talked to. He shrank from society."[26] In a letter, she told her family, "The Hawthornes are invaluable to me, because the little ones come to my room every day and I go there when I like. Mrs. Hawthorne sometimes walks with us, and Mr. H. *never.* He has a horror of sightseeing and of emotion in general."[27]

Although she was not particularly troubled by his aversions—it was clear to everyone that Mr. Hawthorne was a miserable tourist—Hawthorne's dark moods cast a bit of a pall over Mitchell's Roman holiday. She would have laughingly dismissed the possibility of taking an Italian lover (as Fuller had done), but although her conduct was exemplary even by Hawthorne's fastidious standards, there was something about her—almost certainly her intelligence—that smacked of sexual impropriety to at least one of her compatriots. Questions of sex and gender kept coming up in Rome. There was nothing to do but sightsee, and everywhere Mitchell and the Hawthornes went, the sights included goddesses and saints and idealizations of womanhood, from the small, perfect temple of the vestal virgins, to the ubiquitous Virgin Marys, to the classical ruins and museums replete with Minervas and Venuses, to the shop windows filled with reproductions of Beatrice Cenci, and finally to the studios, where painters and sculptors from all over the world worked on their representations of nude women. The nudity shocked Nathaniel Hawthorne, particularly when he found it lifelike. Commenting on the sculptor John Gibson's tinted Venus, he remarked that "this lascivious warmth of hue quite demoralizes the chastity of the marble, and makes one feel ashamed to look at the naked limbs in the company of women."[28] Of *La Fornarina,* Raphael's painting of a topless woman baker, he wrote, "Who can trust the religious sentiment of Raphael, ... after seeing how sensual the artist must have been to paint such a brazen trollop of his own accord, and lovingly?"[29] But brazen trollops were everywhere they looked.

One of the favorite amusements of American tourists in mid-nineteenth-century Rome was visiting sculptors' studios. The community of American artists was tight-knit and highly welcoming to visitors, from whom they hoped to get commissions. There were far more male artists than female, but by 1857, a surprisingly vital community of women artists had started to form. Hawthorne set *The Marble Faun* among this community, making both of his protagonists artists—the delicate Hilda was a copyist, as Sophia Hawthorne had been before her marriage to Nathaniel, and the fascinating but indelicate Miriam was a painter of original works. For Hawthorne, a woman's drive to do original work was itself something of a violation of feminine ideals. *The Marble Faun* praised Hilda for being a copyist rather than an original artist, stating that "she ceased to aim at original achievement in consequence of the very gifts which so exquisitely fitted her to profit by familiarity with the mighty old masters." Perhaps, the novel explained, "this power and depth of appreciation depended partly on Hilda's physical organization, which was ... exquisitely delicate."[30] But the real women artists whom Mitchell and the Hawthornes spent time with were Louisa Lander and Harriet Hosmer, neither of whom could be classified as exquisitely delicate.

Louisa Lander was a sculptor from Salem, Massachusetts. In the first few months of 1858, she made a bust of Nathaniel and was a frequent guest of the Hawthornes. Later, when Mitchell had returned to the United States, the Hawthornes heard a rumor that Lander had posed nude for another artist. They abruptly cut off the connection and were deeply embarrassed by the bust that they had commissioned, which showed Nathaniel's naked collarbones. Hawthorne scholar T. Walter Herbert comments that the exposed collarbones, combined with the rumor about Lander's nudity, "now assumed a potentially lurid meaning, which cast light back on the story of their many hours alone together."[31] Mitchell may have heard the story after the fact; the Hawthornes' very public decision to disown Lander helped end the young woman's career as a sculptor. But although Mitchell mentioned Lander a few times in passing, it was Harriet Hosmer who fascinated her.

Back in the United States, Mitchell wrote an essay about Hosmer that she preserved carefully in her notes. It was saved alongside essays on Galileo and Mary Somerville that seem to be drawn from that same winter, when she explored Galileo's Italy and made the acquaintance of both Hosmer and Somerville. She offered public lectures about these figures in the 1860s, and she probably delivered each lecture many times. In the Mitchell papers that her family carefully saved, the Hosmer essay is collated with the Somerville one as part of a lecture on "Women and Work" and introduced with the words, "I will give you another example of a strong woman to whom no one can refuse admiration"—Mitchell held up Hosmer as a role model for her audiences. In the decades that followed, Hosmer would become well known for her sculptures, particularly her public commission of the statue of the American senator Thomas Hart Benton, her noble *Zenobia in Chains*, and her sleeping *Beatrice Cenci*. Mitchell told the story of meeting Hosmer this way:

> On rainy days, it is all art. There are the cathedrals, the galleries, and the studios of the thousand artists; for every winter there are a thousand artists in Rome.
>
> A rainy day found me in the studio of Paul Akers. As I was looking at some of his models, the studio door opened, and a pretty little girl wearing a jaunty hat and a short jacket, into the pockets of which her

hands were thrust, rushed into the room, seemingly unconscious of the presence of a stranger, began a rattling, all-alive talk with Mr. Akers, of which I caught enough to know that a ride over the Campagna was planned, as I heard Mr. Akers say, "Oh, I won't ride with you—I'm afraid to!" after which he turned to me and introduced Harriet Hosmer.

I was just from old conservative England, and I had been among its most conservative people. I had caught something of its old musty-parchment ideas, and the cricket-like manners of Harriet Hosmer rather troubled me. It took some weeks to get over the impression of her mad-cap ways ... As a general rule, people disappoint you as you know them. To know them better and better is to know more and more weaknesses. Harriet Hosmer parades her weaknesses with the conscious power of one who knows her strength, and who knows you will find her out if worthy of her acquaintance. She makes poor jokes—she's little rude—a good deal eccentric; but she is always *true*.

At the time when I saw her, she was thinking of her statue of Zeno-bia. She was studying the history of Palmyra, reading up on the manners and customs of its people, and examining Eastern relics and costumes.

If she heard that in the sacristy of a certain cathedral, hundreds of miles away, were lying robes of Eastern queens, she mounted her horse and rode to the spot for the sake of learning the lesson they could teach.

Day after day alone in her studio, she studied the subject.... For years after I came home I read the newspapers to see if I could find any notice of the statue of Zenobia; and at length I did see this announce-ment: The statue of Zenobia is on exhibition ... It was after five years. All through those five years, Miss Hosmer had kept her projects steadily turned in this direction.[32]

Clearly, Mitchell admired Hosmer both as an artist and as a persist-ent worker. She was more impressed with the artist's courage, per-sistence, and resilience than with anything else.

Mitchell paid surprisingly close attention in her description to Hosmer's clothes, which were unconventionally androgynous. Mitchell's description of Hosmer's jaunty hat and short jacket makes it clear that she found this attire perfectly acceptable. Nathaniel Hawthorne reacted differently—he described Hosmer as "a small, brisk, wide-awake figure of queer and funny aspect.... She had on petticoats, I think, but I did not look so low, my attention being chiefly drawn to a sort of man's sack of purple or plum-colored broadcloth, into the side pockets of which her hands were thrust."

Still, although Hawthorne found Harriet's mannish jacket disturbing, he did not necessarily condemn her for it: "There never was
anything so jaunty as her movement and action," he declared. "She
was indeed very queer, but she seemed to be her actual self, and
nothing affected or made up; so that, for my part, I gave her full leave
to wear whatever may suit her best, and to behave as her inner
woman prompts.... I shook hands with this frank and pleasant little
woman—if woman she be, as I honestly suppose,—though her
upper half is precisely that of a young man—and took leave, not
without purpose of meeting her again."[33]

Perhaps the most fascinating outcome of the Mitchell and
Hawthorne winter of 1858 in Rome is that all three of them were
keeping careful records. Sophia's account of her first impressions of
Harriet Hosmer echoes and mediates between her husband's and
Maria Mitchell's:

> Miss Hosmer ... came forward with the most animated gesture to greet
> us. Her action was as bright, sprightly, and vivid as that of a bird: a small
> figure, round face, and tiny features, except large eyes; hair short, and
> curling up around a black velvet cap ...; her hands thrust into the pock-
> ets of a close-fitting cloth jacket—a collar and cravat like a man's—and
> a snowy plaited chemisette, like a shirt bosom. I liked her at once, she
> was so frank and cheerful, independent, honest and sincere—wide
> awake, energetic, yet not ungentle.[34]

Both Sophia and Nathaniel Hawthorne liked Hosmer, but
Sophia's reaction is more straightforward than her husband's. Hosmer troubles him—he sees her as worrisomely "queer"—while neither Sophia nor Maria is disturbed.

In 1858, "queer" did not necessarily have sexual connotations; it
was often used as a synonym for "strange." But the word could be
used to describe variation from sexual conventions, and it is clear that
Nathaniel linked Harriet Hosmer's queerness to her complicated
mix of male and female traits and clothing. Nathaniel grudgingly
gave her his permission to continue dressing the way she did, but he
confessed his doubts about her femininity. And yet it would be far
too easy to dismiss Nathaniel Hawthorne as a sexist curmudgeon.
His condemnations of Margaret Fuller and Louisa Lander and his
discomfort with Harriet Hosmer can all be read as the reactionary

stance of a misogynist, although Hawthorne is clearly troubled by women and by women's sexuality. The gallons of ink that he spills in agony over these issues make him, at the very least, an ambivalent misogynist. And *The Marble Faun,* his carefully balanced, strangely ambiguous novel about the independent women artists of Rome, is a heartfelt response to the strange and exciting possibilities for talented women travelers in Rome at the time of its writing.

Mitchell seems not to have dismissed Hawthorne either, but she never spoke about him with the admiration she expressed for Hosmer, the Herschels, or any of the other figures she met on her travels. She parted from Hawthorne as a stranger, treasuring his final note to her, which ended with the hope, "Pray remember me sometimes, when you are not thinking of stars and comets, and believe me, Very Sincerely Yours, Nath Hawthorne."[35] The respectful distance between the two indicates a level of positive feeling; neither was ever afraid to express hostility or disagreement. What is more interesting than their feelings toward each other is the fact that their shared time in Rome immersed them in a shared set of intellectual, philosophical, and deeply personal questions about what it meant to be a woman of genius.

In the end, Rome fulfilled Mitchell's expectations gloriously. She loved to wander in the open fields surrounding the small city and climb over the tumbled stones of the forum. As she wandered the galleries of the Vatican and the Capitol or hiked through the vast expanses of St. Peter's, she often carried a small volume of Byron, just as Margaret Fuller, James Freeman Clarke, and hundreds of other Anglo-American tourists had done before her. The poets, she thought, helped her "to a sense of the beauty of nature and art—a poet seems to have a keener perception." Mitchell gave herself over to the tourist experience, and her best companion in these wanderings was Sophia Hawthorne, who often sketched by her side. Sophia loved being a cultural tourist. In her journal, she repeatedly remarked on her own good fortune, with comments such as "I congratulate myself that I have travelled to Rome from America, if only to see such a consummate work of genius."[36] At the same time, a few feet away, Nathaniel was muttering to Mitchell, "The St. Peter's of my imagination was better!"[37]

On March 16, 1858, Nathaniel Hawthorne stayed home and wrote a letter to a friend. He complained,

> I feel no energy or enterprise, and should really be glad to lie in bed every day, and all day long, if the fleas did not make me so very uncomfortable there.... Mrs. Hawthorne and the children...have gone out for the day (to the Sistine Chapel, I believe,) leaving me by the fireside, which suits me better than any other sight Rome can show. If my pen would but serve me as it has done of yore, I would send you such a description of this cold, rainy, filthy, stinking, rotten, rascally city, as would avenge me for all the incommodities I have suffered here. I hate it worse than any other place in the whole world.[38]

But Sophia Hawthorne wasn't at the Sistine Chapel when Nathaniel wrote that letter. Instead, she wrote in her journal, "It was so perfectly clear and dry and exhilarating, that I took U[na] to the Palatine, to explore the ruins of the Palace of the Caesars.... We strolled about, above, among arches, round towers, chambers, halls and recesses, gathering purple flowers (efflorescent loyalty, in the very home and centre of kingly pomp), and bay leaves, with which to crown Caesar's brow, and ivy and laurestinus—and admiring without end the magnificent views on every side of the lordly Palatine, the Campagna, and the Alban and Sabine hills, whitened with snow—and Rome within these lovely bounds."[39] Mr. and Mrs. Hawthorne had very different Roman experiences. While she wove garlands of flowers and wrote pages and pages of enthusiastic praise of all she saw, he was huddled by the fire, hating the place—and perhaps one of the reasons he was so uncomfortable was precisely that Sophia was so free and happy. She did not hesitate to leave him behind to wander with her daughter and her women friends. Like Mitchell, Sophia was discovering a happy sisterhood in Rome, and Nathaniel didn't like it.

A few years later, in May 1860, when Mitchell was back on Nantucket, she read *The Marble Faun* with great enthusiasm, describing it as a "feast." But although she enjoyed the novel, the author's perspective puzzled her. She wrote to her sister, "Ought not Mr. Hawthorne to be the happiest man alive? He isn't though!"[40]

Mitchell seems to have pitied Nathaniel more than she censured him, as she was inspired and comforted by the very women who made him so gloomy and frustrated. The expatriate Roman sisterhood that peopled Hawthorne's novel sustained and nourished her, even within his shadowed pages. Every tourist who traveled to Rome that winter seems to have published an essay or a book, delivered a few lectures, showed a painting or sold a monumental piece of sculpture recording the experience. Many of the artists and writers were men, but Sophia Hawthorne, Maria Mitchell, Louisa Lander, and Harriet Hosmer, among others, made a formidable sisterhood, and Mitchell's time among them was a source of strength and confidence for the rest of her life.

Rome was not only a locus for thoughtful, cultured Corinnes. Mitchell also thought of Italy as "the very paradise of astronomers," in part because of the remarkable air quality in a region that had not yet felt the stirrings of industrialization.[41] But in Mitchell's mind, even though Italy was the "land of Galileo," Rome was "the city in which he was tried."[42] After publishing a book that argued for the heliocentric solar system—as opposed to the geocentric system propounded by Ptolemy and Aristotle—Galileo was prosecuted by the Church and forced to recant. He did so, but was nonetheless held in house arrest for the rest of his life. Mitchell wrote,

> I know of no sadder picture in the history of science that that of the old man Galileo, worn by a long life of scientific research, weak and feeble, trembling before that tribunal whose frown was torture, and declaring to be false that which he knew to be true. And I know of no picture in the history of religion more weakly pitiable than that of the Holy Church trembling before Galileo, and denouncing him because he found in the Book of Nature truths not stated in their own Book of God—forgetting that the Book of Nature is also a Book of God.[43]

Mitchell loved many of the church buildings of Rome, and felt inspired by the centuries of religious art stored at the Vatican, but she was suspicious of Catholicism, partly because of simple New England Protestant prejudice, but more significantly because she felt that the Church had persecuted Galileo and set itself up as an enemy of astronomical truth. Mitchell later explained Galileo's story in a lecture that she preserved in her files:

> He was tried, condemned, and punished, for declaring that the sun was the center of the system and that the earth moved around it; also that the earth turned on its axis. For teaching this, Galileo was called before the assembled cardinals of Rome, and, clad in black cloth, was compelled to kneel, and to promise never again to teach that the earth moved. . . . He was tried at the hall of Sopre Minerva. . . . It is a very singular fact, but one which seems to show that even in science, "the blood of the martyrs is the seed of the church," that the spot where Galileo was tried is very near the site of the present observatory, to which the pope was very liberal . . . Indeed, if a cardinal should, at the Hall of Sopre Minerva, call out to Secchi, "Watchman, what of the night?" Secchi could hear the question; and no bolder views emanate from any observatory than those which Secchi sends out.[44]

Mitchell does not give the full name of the place where Galileo was tried, which is Santa Maria Sopre Minerva: the church of Saint Mary, built on a former temple of Minerva. Maria and Minerva are two somewhat contradictory icons of Roman womanhood. Mitchell told the story of Galileo's trial at Sopre Minerva in order to introduce her own struggles with the Catholic Church, which banned all women from the observatory run by Angelo Secchi at the Collegio Romano. Mitchell was interested in Secchi's work—he was a pioneer of spectroscopy and one of the first astronomers who could really be called an astrophysicist—but she was not desperate to see his observatory until she learned that women were forbidden to enter it. "I was told that Mrs. Somerville, the most learned woman in Europe, had been denied admission. And that the daughter of Sir John Herschel, in spite of English rank and of the higher stamp of Nature's nobility was at that very time in Rome and could not enter an observatory which was at the same time a monastery. . . . If I had been mildly desirous of visiting the observatory, I was now intensely anxious!"[45]

Mitchell's attitude here shows that her European travels had greatly changed her. No longer a shy Nantucket girl, Mitchell was a force to be reckoned with. She had come to terms with her public stature as the world's foremost woman astronomer and had begun to feel that it was her responsibility to use her position to advance women's access to education. When Mitchell began campaigning to get into the Vatican observatory, her purpose was much more political than astronomical. She didn't need to see another telescope or transit circle; she had read Secchi's writings and spoken to him when he visited her in Rome, and he had already shown her his photographs of Saturn's rings and told her about his work in spectroscopy. No—what she wanted was to fight against the sexist strictures that declared "my heretic feet must not enter the sanctuary, that my woman's robe must not brush the seats of learning."[46] Mitchell appealed to American diplomats and to Church authorities, and finally an acquaintance whose uncle was a monsignor successfully interceded on her behalf. Permission was granted.

To enter the observatory, she walked through the Church of St. Ignasio, "through rows of kneeling worshippers, by the strolling students, and past the lounging tourists" to meet Father Secchi, who waited behind a pillar near the altar. Mitchell's Italian maid (who, Mitchell wrote, "had come to think herself quite an astronomer"[47]) begged to be allowed to accompany them, but Father Secchi demurred, so, as Mitchell described it: "alone, I entered the monastery walls. Through long halls, up winding staircases . . . ; then through the library of the monastery, full of manuscripts on which the monks had worked away their lives; then through the astronomical library, where young astronomers were working away theirs, we reached at length the dome and the telescope. One observatory is so much like another that it does not seem worth while to describe Father Secchi's."[48] She noted perfunctorily that the telescope was about the size of the one in Washington, D.C., and that the Vatican had invested in a sophisticated transit-circle mechanism. There was a bit of historical irony there; as Mitchell put it, "the telescope must keep very accurately the motion of the earth on its axis; and so the papal government furnished nice machinery to keep up with this motion—the same motion for declaring whose existence Galileo suffered! The two hundred years had done their work."

But if the struggle for heliocentrism had been settled long ago, the struggle for women's access to scientific institutions was just beginning. "I should have been glad to stay until dark to look at nebulae," she explained, "but the Father kindly informed me that my permission did not extend beyond the daylight which was fast leaving us, and conducting me to the door, he informed me that I must make my way home alone." After all of the long nights of observation with her own father, the months of hard night work with the men of the Coastal Survey, the hours and hours spent observing with the Bonds at Harvard, and even the nights spent tucked into the tiny makeshift bedroom in the turret of the Greenwich Observatory, this was Mitchell's first experience of being excluded from nighttime observation, because, as a woman, she was seen as a sexual threat. As the back door of the Collegio Romano slammed shut behind her, leaving her unescorted in a dark Roman alley, Mitchell found herself rudely pushed into a battle for female access that she had never imagined she would need to fight. Rome had radicalized her.

Mitchell left Rome in mid-April with a strong feeling of commitment to the fight for women's education and a new awareness that it might, in fact, turn out to be a battle. Nathaniel Hawthorne and Angelo Secchi had been her two great educators in this respect—both were gentle, kind, and fairly sympathetic, but both were genuinely frightened of women such as Mitchell (and Hosmer). Secchi's fears were not personal so much as institutional, while Nathaniel Hawthorne's were deeply and bafflingly personal. Both men were afraid of what Mitchell represented, and Mitchell, who had never in her wildest dreams expected to be a threatening figure, made her way to Florence with much on her mind.

Years later, Mitchell concluded her lecture on Galileo and the Vatican observatory by tying her experience there to the struggles for women's rights.

> We smile at the tho't that a Roman Observatory would not permit a woman to cross its threshold, yet how many boys colleges today are open to women, *except to walk through them?* The Collegio Romano was not alone an observatory or a college, it was also a religious institution, it was a monastery, it was so for them—a church. How many pulpits are

open to the American women today?...Do you know of any case in which a boy's college has offered a Professorship to a woman? Until you do, it is absurd to say that the higher learning is within the reach of American women."[49]

The sense of being excluded was a new and painful one. Oddly, Mitchell experienced it most profoundly at the very time that she was also finding a new sense of connection to a sisterhood of intellectual women. The two experiences may have worked together to raise her consciousness of herself as a political woman allied to the cause of women's educational opportunity.

It was a peculiar fate that brought Nathaniel Hawthorne and Maria Mitchell to Rome in the same carriage, but the winter they spent there, surrounded by a community of intellectual Anglo-American women artists, was formative for both of them. Later, Nathaniel Hawthorne wrote *The Marble Faun* about the mysteriously independent American women artists of Rome, while Mitchell wrote a lecture on Harriet Hosmer's independence that she delivered many times in the following years. Both were fascinated and inspired by the Yankee Corinnes, but Nathaniel Hawthorne was painfully ambivalent about them. After the winter in Rome, Mitchell and the Hawthornes parted ways. Mitchell continued her travels through Europe, while Nathaniel Hawthorne retreated to the hills of Tuscany to start work on *The Marble Faun*. Upon returning from Europe, Mitchell decided to abandon domesticity and throw herself into the intellectual life. Hawthorne retired from his professional life and devoted himself to domesticity.

CHAPTER 8

A Mentor in Florence

Florence was the climax of Maria Mitchell's European tour. Tourists have been falling in love with Florence for centuries, at least since the English poet John Milton's visit in 1638. It is, of course, a stunningly beautiful place, but Milton and Mitchell both loved Florence because it was a city of ideas, welcoming scientists, philosophers, and poets along with artists and curious travelers. A month or so before Mitchell's arrival in Florence in 1858, her friend the Swedish feminist Fredrika Bremer had exclaimed, "Beautiful, blooming Florence ... ! All here is life, movement, beauty! The Arno has cleared its waters, green trees shine forth gaily among the elegant houses, the splendid churches and palaces; marble statues—forms of beauty or pensive thought meet you everywhere, with porticoes and bridges, beneath the blue vault of heaven."[1] Bremer's description captures Florence's rare combination of architectural beauty, intellectual culture, and clear blue sky. Mitchell's interests in astronomy, literature, women, and even progressive religion all came together there. Rome was where Galileo had been condemned, but he had lived in Florence when he had welcomed John Milton and showed him the moon through a telescope. Standing atop the tower at Ascetri, a few miles outside of town, where Galileo and Milton had gazed at the stars together, Mitchell could see the meeting points between the disciplines as clearly as she could see the Tuscan hills. Florence welcomed poets and scientists as Rome welcomed painters and sculptors. The city was stocked with artistic treasures, yet it was also fairly inexpensive, and a large Anglo-American community had taken hold. Florence was not dangerous in the summer (as malarial Rome was) or overwhelmingly damp and cold in the winter (as watery Venice was), so many Anglo-American expatriates made it their permanent home.

Following her experience in Rome, Mitchell had begun to realize that her interest in science would always be linked to an interest in women's issues, simply because she was a woman scientist. Florence helped her to bridge the two worlds, not least because she was finally able to meet Mary Somerville there. Somerville and Caroline Herschel had both been admitted to Royal Society in 1835, so, after Herschel's death, in 1848, Somerville was the only living woman member of the Royal Society—an inspiring parallel for Mitchell, only woman member of the American Academy of Arts and Sciences. Mitchell had high hopes for the meeting, as she had never met a professional woman scientist before.

Most travelers to Florence in the 1850s visited the Museum of Science, which memorialized Galileo. Many noticed his grave, a bit beyond Dante's monument in the Santa Croce cathedral. But with all the treasures of the Uffizi and the Palazzo Vecchio to distract them, not to mention the paintings adorning a multitude of competing churches and monasteries, relatively few tourists made their way out to Galileo's tower home in Ascetri. Mitchell, of course, made a point of it.

The first stop on a nineteenth-century tour of Galileo's Florence was probably the Specola, the new observatory and museum of science that had been built in 1841. Mitchell would have paid a visit to see Galileo's instruments and perhaps to contemplate his index finger, which pointed upward in a small vitrine set into the base of his memorial statue, known as the Tribune. Nathaniel Hawthorne described these rooms after his own visit a few weeks later:

> They consist of a vestibule, a saloon, and a semi-circular tribune, covered with a frescoed dome, beneath which stands a colossal stature of Galileo, long-bearded, and clad, I think, in a student's gown or some voluminous garb of that kind. Around the tribune, beside and behind the statue, are six niches, in one of which is preserved a forefinger of

Galileo, fixed on a sort of little gilt pedestal, and pointing upward, under a glass cover. It is very much shriveled and mummy like, of the color of parchment, and seems to be little more than a finger-bone, with the dry skin or flesh flaking away from it; on the whole, not a very delightful relic, but Galileo used to point heavenward with this finger, and I hope, has gone whither he pointed. Another niche contains two telescopes, wherewith he made some of his discoveries; they are perhaps, a yard long, or rather more, and of very small caliber. Other astronomical instruments are displayed in the glass cases that line the rooms; but I did not understand their use any better than did the monks who wanted to burn Galileo for his heterodoxy about the planetary system.[2]

Sophia Hawthorne's description of the same place started with the observation that "Galileo's heart" had been "thoroughly broken." She wrote, "How little he dreamed, when he sat in prison, that even his fingers would become precious relics for posterity! But I wish he had kept firm, and not denied the truth he had discovered. That is an endless grief to me."[3] Mitchell shared Sophia's grief—her thoughts about Galileo were always tinged with sadness, both because he had been treated badly by the Church and because he had bowed to the pressure. Another visitor to Florence that winter, Fredrika Bremer, imagined the scene more optimistically. When Bremer visited the Tribune memorial, she lamented the fact that there was no picture of Galileo's recantation:

> They have intentionally omitted, amongst the pictured memorials of his life, that moment which is perhaps the most remarkable of all, when in order to free himself from imprisonment in the Romish Inquisition, he denied his assertion that the earth moved around the sun, which the wise fathers in Rome regarded as a contradiction of the doctrine of Scripture—but immediately after the denial he protested against it, and, as if compelled by his genius, stamped upon the earth, and exclaimed, "Ma pur si muove!" (but it turns after all!). What an exquisite subject for a picture.[4]

When Hawthorne visited the Tribune, he found a ghoulishly fascinating desiccated finger and a few cases of instruments that baffled him, and he couldn't resist the contrary impulse to ally himself with the monks. Sophia Hawthorne found evidence of Galileo's despair and reacted with sorrow. Fredrika Bremer thought about his genius and wished that his indomitable spirit had been memorialized somehow. Mitchell left no record of her own thoughts or impressions, though she was the only one of these visitors who had ever handled a telescope. It must have delighted her to verify that Galileo's battered glass was a little smaller than the one she and her father shared.

The Tribune was on a lower floor of the observatory, which was directed by Giovanni Battista Amici and his assistant Giovanni Battista Donati. Perhaps Mitchell was tiring of observatory tours, because her journals make no mention of meeting these astronomers. A few weeks after she left Florence, Donati would discover one of the most spectacular comets of the nineteenth century, so large and bright and easy to see that it would feature in many mid-nineteenth-century lithographs and paintings. Even Nathaniel Hawthorne was entranced by it. "This evening, I have been on the tower top, star gazing and looking at the comet, which waves along the sky like an immense feather of flame."[5] Intent on her research on Galileo and Milton, perhaps busy with her calculations for the almanac, and trying to make the most of the short time when she could visit Mary Somerville, Mitchell might have stayed away from the Florence observatory altogether.

As early as 1854, Mitchell had mentioned in her journal that she was working on a lecture on Milton and astronomy. She thought her dual insights into poetry and astronomy might help her to shape a new, less cloistered career. She and Nathaniel Hawthorne had searched together in Rome for the house where Milton had stayed, and when her essay "The Astronomical Science of Milton" was finally published, after her death, it began with a description of Galileo's tower outside Florence:

> I visited the Tower of Galileo, on the hill of Ascetri, the same from which he made his observations when Milton was with him. It is only a short and a most lovely drive from Florence. The tower is in pretty good preservation, and we ascended to the top. It must have been a fine place for sweeping the sky with a little glass; the position is elevated, and nothing for miles around but the lovely scenery, and above, the clear sky of Tuscany.[6]

Mitchell had no telescope of her own, and she had not borrowed one for her European trip. She could not actually emulate Galileo's observations, but her purposes were more personal than astronomical.

Mitchell thought of history as progressive, as did most of her peers in the optimistic nineteenth century. She genuinely believed that humankind was making forward progress, and that human knowledge and the human condition continued to improve. For her, both Galileo and Milton were exemplary figures in the progress of humanity. She once wrote that she found it almost impossible to imagine speaking English without quoting the latter: "Milton, when read in childhood, fastens his Heaven and Hell upon us; we cannot forget them,—we know no other. We see no sunrise without thinking of his lines . . . we could not do without the common quotations."[7] But Milton was more than a cultural cornerstone for her; he was also a symbol of progress. Mitchell saw Milton's heaven and hell as improvements upon Dante's, just as she saw Protestant Christianity as a step forward from Catholicism. She praised the poet for being "vastly beyond his age in most respects," though she had a few cavils with him.[8] Of these, the most significant was her conviction that Milton disrespected Eve's intelligence. "As a woman," Mitchell wrote,

> I do not like Book VII [of *Paradise Lost*]. I felt, even when a child, indignant that Milton should represent Eve as so careless of the angel's discourse that she must tend her flowers just at that juncture. . . . It seems to me that the childlike Eve should have remained and listened, asked questions, and kept up the dramatic interest. The educationists of today would scarcely be willing to say to an inquiring child, "Solicit not thy thoughts with matters hid."[9]

Mitchell also commented archly that neither Adam nor the angel is "very scientific"—in fact, the angel is downright "ignorant," shuffling and noncommittal. "Adam expresses himself fully satisfied" with the angel's explanation of the cosmos, Mitchell tartly remarks, "which is more than I am and I suspect more than Eve would have been."[10]

Although Mitchell was irritated by Milton's sexism, her understanding of human progress allowed her to tolerate it. She believed

that the world was getting fairer and that opportunities were naturally more extensive in her age than in previous ones. Women's access to education (and to an angel's discourse) might have been
limited in Milton's time, but Mitchell thought that his poetry had
moved humanity toward an enlightened age when educational opportunities would be open to all. She valued him for his language
first and his vision second.

Milton was as interested in the shifting scales of infinity as
Mitchell herself, and he was the first and perhaps the greatest poet of
Copernican astronomy. In Florence, Mitchell concentrated on Mil-
ton's descriptions of Galileo's moon:

> The Moon, whose Orb
> Through Optic Glass the Tuscan Artist views
> At Ev'ning from the top of Fesole,
> Or in Val d'Arno, to descry new Lands,
> Rivers or Mountains in her spotty Globe.[11]
> . . .
> As when by night the Glass
> Of Galileo, less assur'd, observes
> Imagin'd Lands and Regions in the Moon.[12]

What Mitchell appreciated most in Milton was his ability to describe
the shifting perspectives of the newly sun-centered astronomy and
his conviction that "the new Cosmogony" (as Mitchell called it)
could be understood both spiritually and scientifically.[13] Milton had
understood the philosophical implications of the shift from an earth-
centered model of the universe to the idea of a solar system, and he
was alive to the revelatory implications of the telescope, which
helped to indicate the vastness of space and the relative smallness of
the earth. She never thought of Milton as an astronomer, but she valued his "poet's lens"[14] almost as much as he valued "the Glass of Galileo."[15] Milton brought astronomy and theology together within a
poetic frame, and Mitchell loved him for it.

But if Milton was a great figure for Mitchell, Galileo was even
greater. He was the astronomer who first used a telescope to look at
the sky. As Mitchell put it, "Galileo had made known the existence
of the satellites of Jupiter, the belts of Saturn, the inequalities of the
moon's surface, and had declared with fear and trembling, which
time showed to be well grounded, the motion of the earth."[16] Of

course, his significance went beyond these discoveries. For Mitchell, Galileo was "not a mere observer and discoverer, but a philosopher"[17] who carefully negotiated the emerging, crucially important gaps between science and faith.

When Galileo was forced to recant his position, science and religion were set at odds with each other. The fact that he reluctantly agreed shows that Galileo cared deeply about his faith, but he also knew that no public disavowal would stop the earth from moving. As Mitchell wrote, "It is said that when he arose, he whispered, 'It does move!'"[18] The difficulty of reconciling scientific truth with Christian tradition was ever present to Mitchell too, but she genuinely believed that liberal Christianity provided strategies for embracing both spiritual and scientific truths. It was extremely important to Mitchell that the heaven and hell that shaped her understanding of spirituality were framed within a scientifically accurate cosmos.

Mitchell saw herself as a member of a vanguard similar to Galileo's, but rather than argue for the motion of the earth, she would fight for the mobility of women. In that sense, Elizabeth Barrett Browning's *Aurora Leigh* had done more to shape Mitchell's sense of human progress than *Paradise Lost* had. As much as Mitchell loved Milton, she shared Barrett Browning's more contemporary views of women. In *Aurora Leigh,* Barrett Browning writes:

> But poets should
> Exert a double vision, should have eyes
> To see near things as comprehensively
> As if afar they took their point of sight,
> And distant things as intimately deep
> As if they touched them. Let us strive for this.
> I do distrust the poet who discerns
> No character or glory in his times.[19]

Browning's telescopic sense of vision in these lines is similar to Milton's sense of shifting perspective when he adopts the Galilean view. Another similarity between the poets is that both considered themselves reformers: Milton was a devoted Puritan at a time when Puritanism was radical, while Barrett Browning's *Aurora Leigh* described nineteenth-century reform movements in the age of Marx and Fuller. The conviction expressed in the last lines of *Aurora Leigh* perfectly expressed Mitchell's beliefs in historical progress:

> The world's old,
> But the old world waits the time to be renewed,
> Toward which, new hearts in individual growth
> Must quicken and increase to multitude
> In new dynasties of the race of men;
> Developed whence, shall grow spontaneously
> New churches, new oeconomies, new laws
> Admitting freedom, new societies
> Excluding falsehood.[20]

Just as Milton had embraced Galileo's heliocentric vision and re-arranged his account of paradise to accommodate scientific progress, Barrett Browning was a poet of her time, picking up on the evolutionary narratives that were making nineteenth-century thinkers such acolytes of progress. Neither Mitchell nor Barrett Browning had read *The Origin of Species*—it was published in 1859, the year after Mitchell's visit to Florence—but evolutionary narratives, social and otherwise, were part and parcel of the spirit of the age.

As it turned out, Mitchell never met Elizabeth Barrett Browning in Florence. Later, she told the story to her students, according to Mary King Babbitt:

> One day, at the beginning of the recitation period, she told us the story of how she did not meet Mrs. Browning. When she was in England, a well-meaning friend of both women gave her a book belonging to Mrs. Browning, asking her to return it to the owner. Miss Mitchell left it at Mrs. Browning's door, ignoring her friend's little scheme to bring about an informal meeting, returned to America, and never saw the poet. "And this, girls," she said in conclusion, "is what I should wish you to do under similar circumstances. Never thrust yourselves upon anyone who does not surely want you."

Babbitt went on to comment, "This hearty respect for individuality was perhaps the keynote to her character."[21]

Mitchell may have regretted her failure to cross the threshold of the Brownings' Casa Guidi, even though she was proud of her refusal to become another "pushing American." But she did not have much time in Florence, and she could not waste any of it on regrets; there was a much more important visit to make. She carried a letter of introduction to Mary Somerville.

Maria Mitchell was desperate to meet Mary Somerville. As she had planned her trip to Europe, she had asked every astronomer and mathematician she knew for a letter of introduction. Through her work for the almanac, she knew most of the prominent Americans in both fields, yet none had come through. In England, she continued to try to find a helpful conduit. In view of Mitchell's fastidious hesitancy where Elizabeth Barrett Browning was concerned, her willingness to ask relative strangers for a letter to Somerville shows the enormous importance she attached to this meeting.

Mitchell thought of both Barrett Browning and Somerville as "women of genius,"[22] but of course she and Somerville toiled in the same professional field. William Whewell had coined the word "scientist"—to replace the more common early-nineteenth-century terms "natural philosopher" and "man of science"—in a review of one of Mrs. Somerville's books. Somerville was literally the first professional scientist to appear in the chronicles of nineteenth-century thought.

Whewell had not treated Mitchell with much respect, but he would never have questioned her esteem for Mary Somerville. On the contrary, he was instrumental in the decision to adopt Somerville's book on LaPlace's astronomy as a requirement for Cambridge students specializing in higher mathematics. Somerville had published her translation of Pierre-Simon LaPlace's *Mechanism of the Heavens* in 1831, and her work was universally acknowledged as more than a translation: she had improved upon LaPlace's calculations for the sake of clarity, producing a work more lucid and elegant than its notoriously difficult original.

Translation was a perfect entry into scientific discourse for Somerville, who had no formal education and limited access to the mathematical and scientific communities. Under the guise of being a mere translator, she proved herself competent at the most difficult calculations of her era. LaPlace was delighted with her translation. According to Mitchell, LaPlace was unaware that Somerville had changed her name when she remarried after her first husband died. When the two met, he remarked, "Only two women have ever read the 'Mecanique Celeste'; both are Scotchwomen—Mrs. Greig and yourself."[23] Somerville was Mrs. Greig, of course, and she was in a class by herself. Although others could read LaPlace in the original, no one could match Somerville's ability to clarify his mathematics. Whewell had gone so far as to write poetry on Somerville's genius:

> ...dark to you seems bright, perplexed seems plain,
> Seen in the depths of a pellucid mind,
> Full of clear thought; free from the ill and vain
> That cloud our inward light.[24]

These lines are from a sonnet published in Whewell's review of Somerville's second book, *On the Connexion of the Physical Sciences* (1834)—the book that made Somerville so significant in the history of science. When Whewell groped for words and finally coined "scientist" to describe her, the issue was not primarily gender, but rather the newness of Somerville's endeavor—her attempt to connect all the physical sciences to one another. Whewell lamented, "Physical science itself is endlessly subdivided...the mathematician turns away from the chemist; the chemist from the naturalist; the mathematician, left to himself, divides himself into a pure mathematician and a mixed mathematician, who soon part company," and so on. Because of these disciplinary divisions, Whewell explains, "physical science loses all traces of unity. A curious illustration of this result may be observed in the want of any name by which we can designate the students of the material world collectively." "Philosopher" was inadequate, he explained, because "it was felt to be too wide and too lofty a term, and was very properly forbidden" by none other than the poet Samuel Taylor Coleridge. And so Whewell proposed the new word "scientist."[25]

Another, even more important reason that Whewell (and Coleridge) felt the need for a new term was that a new professional identity was developing. Those who studied the material world were beginning to distinguish themselves from philosophers, whose provinces were more metaphysical than physical. But the first steps of this separation had been quite insulated from each other: chemists, mathematicians, astronomers, and the soon-to-be-named physicists did not necessarily see themselves as sharing an identity or as working at a common endeavor. Somerville's treatise *On the Connexion of the Physical Sciences* was instrumental in showing the various investigators that their work was connected—they were all practitioners of science.

Although the development of the word "scientist" related more to the philosophical point (argued by Somerville) that the sciences could be unified than it did to gender, "scientist" did gradually replace the older formulation "man of science." Gender also entered in, Whewell thought, because as a woman, Somerville was better equipped to see connections than a man. Whewell firmly believed, "Notwithstanding all the dreams of theorists, there is a sex in minds," and further that "when women are philosophers, they are likely to be lucid ones; that when they extend the range of their speculative views, there will be a peculiar illumination thrown over the prospect."[26] Whewell argued that Somerville's womanly perspective enhanced rather than obscured her vision, and allowed her, in turn, to illuminate the connections to men mystified by their own warring impulses, caught in "inextricable confusion—an endless seesaw of demand and evasion."[27]

Somerville corresponded with John Herschel, Michael Faraday, and William Whewell and was familiar with the work of almost every scientist in Europe. Her talent for synthesis made her perhaps the most widely informed scientist of her generation. Even William Whewell (who made a great effort to keep his finger on science's pulse) was in awe of Somerville's ability to stay on top of scientific developments: "Those who have most sedulously followed the track of modern discoveries cannot but be struck with admiration at the way in which the survey is brought up to the present day. The writer has 'read up to Saturday night,' ...; and the latest experiments and speculations in every part of Europe are referred to."[28] Her later works, *Physical Geography* (1848) and *On Molecular and Microscopic Science* (1869), continued her efforts to unify the nineteenth-century sciences.

Whewell closed his review of Somerville's work with a verse modeled on John Dryden's "Lines on Milton," but replacing Milton with Somerville. Dryden's poem begins:

> Three poets, in three distant ages born
> Greece, Italy and England did adorn . . .[29]

In Dryden's original poem, the Greek poet is Homer, the Italian poet Virgil, and the English poet Milton. Shifting the lens, Whewell's poem refers to Hypatia of Alexandria, Maria Agnesi of Bologna, and Mary Somerville:

> Three women, in three different ages born,
> Greece, Italy, and England did adorn;
> Rare as poetic minds of master flights,
> Three only rose to science' lofty heights.
> The first a crowd in brutal pieces tore,
> Envious of fame, bewildered at her lore;
> The next through tints of darkening shadow passed,
> Lost in the azure sisterhood at last;
> Equal to these, the third, and happier far,
> Cheerful though wise, though learned, popular,
> Liked by the many, valued by the few,
> Instructs the world, yet dubbed by none a Blue.[30]

Whewell's reference to Hypatia's death is fairly clear, though his description of Agnesi's decision to enter a convent is a little more obscure ("lost in the azure sisterhood"). His final point is that Mary Somerville is not a tragic figure, as Hypatia was, or a celibate Catholic nun, as Agnesi became. Instead, England had managed to produce a happy woman scientist—so conventionally feminine that her greatest triumph is that no one thinks of her as a bluestocking ("dubbed by none a Blue"). Perhaps the most striking thing about the verse, however, is that Whewell intentionally likens Somerville to Milton by closely echoing Dryden's poem. This was no mere coincidence; Kathryn Neeley, Somerville's twenty-first-century biographer, argues that Somerville "took the poetic traditions established by Milton . . . and transformed them for scientific prose."[31] Like Milton, and indeed like William Whewell and John Herschel, two of her closest correspondents, Somerville's greatest contribution to science was her writing. Her books helped create the nineteenth-century sciences, and helped the very first "scientists" to recognize themselves as such.

Whewell's review of Somerville's work concluded by saying that "we are very happy to claim her as one of the brightest ornaments of England."[32] In 1835 she was elected to the Royal Astronomical Society, and shortly thereafter the larger Royal Society "voted unanimously to have her bust placed in their Great Hall."[33] Perhaps most substantially, in 1835 she was granted a pension on the British civil list. Robert Peel, the Tory prime minister who recommended her to the king for a pension, notably told Somerville that its purpose was "to encourage others to follow the bright example which you have set, and to prove that great scientific attainments are recognized among public claims"[34]—not merely to recognize Somerville individually, but also to connect the state with scientific achievement. John Herschel, William Whewell, and many other fellows of the Royal Society were jubilant about this recognition because it "constituted a step in the professionalization and institutionalization of science."[35] The parallels between Mitchell's comet medal and Somerville's pension are clear—both women were among the first government-supported professional scientists in their respective countries, and both women's individual achievements helped to establish scientific institutions. By the time she died in 1872, Somerville had reigned for nearly fifty years as the "queen of science," according to her obituary in the *Morning Post*. In 1879, less than a decade after her death, Somerville Hall, the second women's college at Oxford University, opened its doors.

Mary Somerville won many accolades for science, but she was equally celebrated for her conventional femininity. In 1832, Maria Edgeworth said she admired Somerville because she was "the only woman who understands" LaPlace's astronomy, "and while her head is up among the stars, her feet are firm upon the earth."[36] For the rest of the century, praise of Somerville would continue to ring, following Edgeworth's melody. As often as Somerville was acknowledged to be the foremost woman mathematician of the age, she was also praised for her personal modesty and quiet charm. Kathryn Neeley sees Somerville's modesty as a key to her success, and Maria Mitchell agreed. Mitchell's own essay on Somerville in the *Atlantic Monthly* concluded by saying,

No one can make the acquaintance of this remarkable woman without increasing admiration for her. The ascent of the steep and rugged path of science has not unfitted her for the drawing room circle; the hours of devotion to close study have not been incompatible with the duties of the wife and mother; the mind that has turned to rigid demonstration has not thereby lost its faith in the truths which figures will not prove.[37]

Appropriately enough, it was a network of women who finally gave Mitchell what she longed for: Richarda Airy, the wife of the Royal Astronomer at Greenwich, introduced Mitchell to Lady Margaret Herschel, Sir John Herschel's wife, and Lady Herschel gave Mitchell a letter of introduction to Mary Somerville. Mitchell's account makes it clear that meeting Somerville was even more important to her than meeting the Herschels: "The Airys were just and kind to me; the Herschels were lavish, and they offered me a letter to Mrs. Somerville." It was an "open sesame to Mrs. Somerville's heart," Mitchell fondly recalled in an essay she published upon her return.[38]

Clutching John Herschel's letter of introduction, Mitchell presented herself at Somerville's residence in Florence. She was warmly welcomed, in part because Mrs. Somerville knew of her comet discovery. Later, Maria told her friend Julia Ward Howe that Somerville spoke "with a strong Scotch accent, and said to me 'Ye have done yeself great credit' "[39]—high praise from the woman Mitchell most longed to emulate. Immediately, the women plunged into exactly the sort of scientific discussion for which Mitchell was starved after almost a year away from the more egalitarian American astronomical community. At their first meeting, Mrs. Somerville's conversation ranged over many topics. Mitchell recalled that

she touched upon recent discoveries in chemistry or the discovery of gold in California, of the nebulae, more and more of which she thought might be resolved, and yet that there might exist nebulous matters, such as compose the tails of comets, of the satellites, of the planets, the last of which she thought had other uses than as subordinates. She spoke with disapprobation of Dr. Whewell's attempt to prove that our planet was the only one inhabited by reasoning beings; she believed that a higher order of beings than ourselves might people them.

On subsequent visits there were many questions from Mrs. Somerville in regard to the progress of science in America. She regretted, she said, that she knew so little of what was done in our Country.[40]

Mitchell visited Somerville at least three times over the brief week she spent in Florence. During that time, Somerville also told Mitchell about the progress of her manuscripts: "She told me that her new edition of the 'Physical Geography' was now in press in London, but that of the 'Physical Sciences' she had not yet received the proof sheets."[41] After a later visit, on April 20, Mitchell recorded, "Mrs. Somerville told me that an English gentleman named Joule had advanced the idea of late that heat is motion and that she has enlarged upon this in her book and she gave me various anecdotes illustrative of this doctrine. She remarked also that the science of magnetism had made strides in the last few years."[42] Finally, at their last meeting, Mitchell joyfully recorded, "Mrs. Somerville met me as I entered, and told me that she was glad to see me and that she had expected me."[43]

Sadly, Mitchell was only able to spend a week in Florence. Time and money were both running out, and she had already planned the arduous journey over the Alps toward Berlin. Her meetings with Somerville were plenty for a single week, but not so many for a lifetime. Even in such brief encounters, Mary Somerville was able to act as a mentor to Mitchell in a way that no other astronomer or scientist ever had. Somerville talked to her about substantive scientific questions as none of the British scientists had done; Mitchell first learned about the work of the physicist James Prescott Joule in Florence, despite having spent months in scientific circles in England, where Joule lived and worked. Somerville took Mitchell seriously as an intellect, and wanted to share her wide-ranging knowledge and encourage Mitchell in her own endeavors. She made her affection for Mitchell clear, and she offered the support and encouragement the younger scientist needed. Best of all, Mitchell liked her. She was charming and kind, someone for Mitchell to emulate in every way.

Mitchell had published a few scientific papers before her journey, but her first mainstream publication would be an essay on Mary

Somerville, published in the *Atlantic Monthly* in May 1860. She wanted to spread the word because she believed that Somerville was the exact sort of role model needed by women who were interested in science. Mitchell now believed it was more possible in the nineteenth century than it had ever been for women to be intelligent, successful, and happy.

What Mitchell admired most in Somerville was her ability to imagine the future of science. As Mitchell put it, Somerville could not only develop the "theory of worlds" and trace back their history; she could also "sketch the outline of their destiny."[44] A true visionary, Somerville saw an incipient world that provided opportunities and encouragement to scientific women. Together, Somerville and Mitchell discussed astrophysics, nebulae, geology, geography, and politics, but what Mitchell finally gained from Mary Somerville was a sense of mission. She came to believe that science needed women precisely because of the unique contributions they could make, and that women scientists needed other women scientists to help them fulfill their potential. After only a week of Mary Somerville's friendship, Maria Mitchell knew that she must spend the rest of her life encouraging younger women scientists, just as Somerville had encouraged her. Mary Somerville would be a touchstone for Mitchell for the rest of her life.

CHAPTER 9

The War Years

When Mitchell came home to Nantucket in the summer of 1858, the island was sadly diminished. Whale ships rarely used the harbor, preferring the access to the railroads at New Bedford. Many families had left altogether, lured toward the gold fields of California or the burgeoning cities along the coast of the mainland. The sandy streets were quiet. Outlying houses stood abandoned. The lobby of the Pacific Bank was deserted, and upstairs in the apartment that a dozen Mitchells had shared from 1837 to 1857, most of the rooms were empty. Maria's mother, Lydia, kept to her bed most of the time. Her father, William, had little to do downstairs at the bank, since the remaining Nantucketers didn't have much money to deposit with him. Her brothers and sisters had all married, and most had moved away; only Kate and Annie remained on the island.

The depopulation of the Pacific Bank building at 61 Main Street was just one example of the exodus from the island. After reaching a population of about 9,000 people from 1840 to 1850, at the height of the whaling boom, the island's fortunes and population declined precipitously for twenty-five years straight. In 1860, there were only 6,000 residents on the island. By 1875 the population would be about 3,000.[1] Nantucket had always been isolated, but in the late 1850s, it was sinking into desolation.

Mitchell moved into a corner of her mother's sickroom and did her calculations for the almanac as she sat near her. She read novels and poetry, wrote letters to her far-flung sisters and her old Nantucket friends, kept up with scientific journals, and read newspapers and magazines, which were increasingly obsessed with the looming war. The contrast between 1858 and 1859 was sharp: Mitchell now spent her days in her ailing mother's bedroom in a large and lonely apartment above a failing bank in a deserted town on an isolated island. A year before, she had been sipping coffee at the Caffe Greco, eating gelato at Spillman's, wandering the museums and the ruins with Sophia Hawthorne, and visiting the studios of leading American artists, from Louisa Lander to Harriet Hosmer to Paul Akers. Back on Nantucket, Mitchell contemplated her European experiences and worked on revising her journals into narratives, sharing them privately with her family and wondering how to make her ideas and experiences more public. By the end of the winter, Mitchell had started drafting her essay on Mary Somerville.

The three years from July 1858 to July 1861 were among the grimmest of Mitchell's life. In a very real sense, she was waiting for her mother to die. She dreaded the death, but she also found it difficult to face her mother's painfully empty life. Afflicted with dementia and unable to understand her own condition, Lydia Mitchell wandered in a maze of unhappy confusion that had changed her completely. The intelligent, well-organized mother with whom Mitchell had shared passions for literature and for household order was almost completely gone.

Altogether, Lydia's decline lasted for six years. For the first two years, from 1855 to 1857, Maria attended to her by herself. In 1857, when her mother's condition stabilized—with no sign of decline but no hope for recovery—one of Mitchell's married sisters agreed to move back in and watch over their mother so that Maria could make her European trip. Upon her return, Maria went back to her nursing duties. Later, her sister Phebe remarked about her mother that "it used to be said that Maria's eyes were always upon her."[2] This was only partly true: Mitchell often turned her eyes to the night sky.

Although these were sad and difficult years for her, Mitchell does not seem to have been caught up in the same sort of self-doubt that had filled her from 1855 to 1857. She did not keep a journal during this time but instead began to use her notebooks for brief working notes. She was past wondering who she wanted to be or what she wanted to do with her life, and she had happily resolved her questions about public and private womanhood. She would not return to the sheltering quiet of the Atheneum; she had outgrown it. Instead, she would continue her work for the United States government as computer of Venus, and she would work as a full-time astronomer with the goal of publishing her results in national journals. Maria Mitchell had become a professional.

One of the hallmarks of a professional scientist is publication. When Mitchell discovered the comet in 1847, the notices about it were published under her father's name. It was not until after her return from Europe that Mitchell would embark on publishing in her own name. The *Atlantic Monthly,* though prestigious and influential, was not a scientific journal, and Mitchell's 1860 piece on Mary Somerville was not a work of scientific research. Still, it was an important step toward becoming a publishing scholar. Mitchell was steeling herself to begin publishing her own scientific work under her own name.

Later in 1860 she wrote triumphantly to her sister, enclosing a copy of an astronomical article she had published in the *Nantucket Inquirer.* "I send you a notice of an occultation; the last sentence and the last figures are mine. You and I can never occult, for have we not always helped one another to shine?" Pleased with her own glimmer of brilliance, Mitchell's letter framed it in the context of sisterhood, deflecting the spotlight onto the sister who had helped her to shine.

It is notable that Mitchell was never discouraged from publishing her work because she was a woman. To the contrary, she received much encouragement. Her supervisor at the almanac urged her forward, writing: "I know that you are one of those who have discovered that the true secret of happiness lies in occupation. . . . I am glad to find that your calculations have not ended in smoke. In respect to . . . the productions of your precious brains, you must not be meek and lowly with these."[3] In 1860 and 1861, along with her computing work and her incessant care for her mother, Mitchell published four or five brief scientific articles in the *Nantucket Inquirer* "on the eclipse of 1860 and the occultation of mars, a huge sunspot, a brilliant meteor, [and] a globular comet," according to Helen Wright.[4] The transition to being a publishing scientist was difficult for her—she overcame fear and much reticence in order to do it. But as a professional scientist, she couldn't help comparing herself to the women of genius whom she had admired so much in Europe, particularly Mary Somerville and Mrs. Browning. Regardless of her public reputation, she felt obliged to publish because she knew she saw "so many beauties" that must be shared with other astronomers and other women. By 1863 she would publish her first solo article in *Silliman's Journal,* the premier scientific journal of the period.

Mitchell's professionalization took place during a period of intense social change. The middle decades of the nineteenth century were among the most tumultuous in U.S. history, and the Civil War epitomized the uproar. Like all wars, the conflict was chaotic and violent, but more than most, it was a symptom of widespread social upheaval. Some social changes centered on race: slavery was abolished in 1862, and the vote was extended to African American men in 1870. Women, white and black, played a large role in the abolitionist movement, and women's political and social positions during this time were being profoundly redefined. The underlying socioeconomic structure of the United States had changed radically in the first half of the century: the United States was now definitively an industrial rather than agricultural nation, and one of the most significant shifts of the period had been the move toward a wage-based economy. Although women and African Americans were often paid much less for their labor than white men, the chance to be paid at all was a huge innovation, granting economic and social mobility that would have been unimaginable when the only road to wealth was owning the whole farm.

But while there were broad increases in mobility and opportunity for many people, the middle of the nineteenth century also saw a surge of domestic ideology that demonized women's mobility in particular, and valorized middle-class women who stayed home in their own parlors. In reaction to the new age of more varied, mixed, and changing social circles than had ever been historically possible, many people celebrated the imaginary space of a quiet, stable, homogenous home life.

The world of science reflected these larger changes: during the middle of the nineteenth century, science in the United States was rapidly professionalizing, as scientific workers were increasingly legitimized by their wages. Scientific institutions that had started as private social clubs were fading away, while government- and university-sponsored scientific laboratories and institutions were gaining prominence. The science of the parlor, where girls and boys, women and men discussed botany and astronomy as a shared interest, no longer looked much like science. Instead, real science was being defined as an activity that could only happen inside the ivy-clad walls of a professional institution.

These myriad social changes happened at the same time, and for someone such as Mitchell, they were all tangled together. The wage economy was an incredible boon to her; she earned significant amounts of money from her work for the almanac, and she was genuinely independent because of it. As one of the very few scientists who had secured a professional job at the very beginning of the move toward professionalization, she was secure in her identity. But she had few hopes of affiliating with a college or university, since most of them were closed to women. A few decades earlier, this would have been a minor concern, since science was not confined to university campuses, but now it was an intractable problem. Mitchell's full professionalization would have been impossible had it not been for the "women of America." Spearheaded by Sophia Hawthorne's remarkable sister Elizabeth Peabody, the women of America contributed for years to a fund to buy a world-class equatorial telescope for Mitchell. In July 1857, just as Mitchell was leaving for Europe, *Emerson's United States Magazine* had published an article about the fund-raising effort. Beneath the headline "Maria Mitchell," it said:

> A Boston friend writes us that the ladies of that city have it in contemplation to start a subscription paper, for the purpose of raising three thousand dollars, to purchase a telescope for this distinguished and truly noble woman, who has devoted herself with so much zeal to the pursuit of science. This sum will purchase an instrument much larger than the one now owned by Miss Mitchell, and will thus greatly facilitate her in her studies.
>
> We sincerely hope something of the kind will be done, and it will be a most womanly tribute to one of the most gifted an deserving of her sex. In Europe, Maria Mitchell would command the interest and receive the homage of the learned and polite, while in America so little prestige is attached to genius or learning that she is relatively unknown. This is a

great fault in our social aspect, one which excites the animadversion of foreigners at once. "Where are your distinguished women—where your learned men?" they ask, as they are invited into our ostentatiously furnished houses to find a group of giggling girls and boys, or commonplace men and women, who do nothing but dance, or yawn about till supper is announced. We need a reform here, most especially if we would not see American society utterly contemptible.[5]

The whole time Mitchell had spent in Europe, she had been planning her new observatory. She decided to spend most of her money on the telescope itself: a five-inch refractor made by Alvan Clark, the best telescope maker in the United States. She sprang for an equatorial mounting, the kind she had seen in use in every observatory in Europe. The trustees of the Coffin School allowed her to sink a granite pier into a plot of land just behind the school, and she constructed upon it a simple observatory with a dome that rotated on ball bearings made of cannonballs.

Mitchell's observatory was small—eleven feet around and only about eleven feet tall—but she was no mere backyard astronomer. When the observatory was finally set up in October 1859, *Scientific American* reported on it:

MISS MITCHELL'S TELESCOPE AND OBSERVATORY

The mechanism of this telescope is probably not excelled by those manufactured in Europe. Its focal strength is between five and six feet, and the clear aperture of the object glass is five inches. It is mounted equatorially.... The telescope plays upon a stand, composed of slabs of iron intersecting each other at right angles, and resting on four points adjusted by foot-screws. This rests on a pier of solid masonry, surmounting a mass of granite. The base of the pier is laid so low that the severest frost cannot affect it, and the whole is isolated from the surrounding earth by beach sand, securing it from the tremor to which it would otherwise be exposed from passing carriages on a road, unpaved at a distance of fifty yards. The great Russian astronomical observer Struve has said that an observatory should be a covering for a telescope. Either from this precept, or from necessary economy, Miss Mitchell has constructed her observatory in such a manner as merely to shelter the instrument. On entering it, however, it is found to possess all the needful equipments of a more costly establishment. It is a circular building,

eleven feet in diameter and scarcely more in height, covered with an ordinary roof made to revolve on cannon balls. By this means, a narrow
aperture in the roof is easily brought to the point of heavens under in-
spection. As the chief object of Miss Mitchell is to devote the instrument to scientific work, the whole period since its construction has
been employed in making those nice adjustments so necessary to useful
results.[6]

The observatory was on land attached to the Coffin School,
which gave it a small degree of institutional affiliation: Mitchell's
brother-in-law Alfred Macy was the principal of the small academy
in the beautiful brick Greek Revival temple on Winter Street, and
Mitchell welcomed the boys and girls who attended the school to
peer through her telescope at every opportunity. She did not offer
formal instruction in astronomy, but she did inspire lifelong interest
in the stars and admiration for herself in the students who had the
chance to visit her there.[7]

If Mitchell had been asked in 1859 to predict her future or explain her hopes for her career, her response would have centered on
the new observatory on Winter Street. She had corresponded with
her European astronomer friends, and with Admiral Smyth's advice,
she was setting out on a study on binary stars. That July, Smyth had
written to congratulate her: "we are much pleased to hear of your
acquisition of an equatorial instrument under a revolving roof, for it
is a true scientific luxury as well as an efficient implement."[8]

But on October 16, 1859, just as Mitchell was starting work in
her new observatory, the abolitionist John Brown raided the United
States Armory at Harper's Ferry. Later, Herman Melville would call
Brown "the meteor of the war." Brown had a wild beard trailing out
from a passionate face, like the fiery tail of a shooting star; his attempt to start an uprising against slavery was seen by most northern-
ers as a portent more ominous than any meteor. It was becoming
devastatingly clear to all Americans that the political differences over
slavery could not be resolved without violence.

On Nantucket, the approaching war affected everyone. Most
Nantucket Quakers opposed slavery, but the early years of the nine-
teenth century had divided the Friends' meetings all across the
United States: some Quakers found themselves compelled by reform

movements such as abolition and women's rights, while others were more interested in tradition. Another long-standing tenet of the Quakers was nonviolence. Nantucket Quakers had always had problems with pacifism, and many had served in the Revolutionary Army and again during the War of 1812, but the Civil War was different— it would be fought in large part in order to abolish slavery, a cause that was central to many Quaker's beliefs. Nonviolence and antislavery arguments collided in the context of a society that was already economically unstable because of the failing whale fishery and the troubled banking system. The town spun apart. If war came, it would be almost impossible to maintain the place as an independent community. Residents were wholly dependent on shipping; if the American coast became a war zone, life in Nantucket would be untenable. From the moment of John Brown's raid on Harper's Ferry, the writing was on the wall for Nantucket. Most islanders made plans to leave.

If the war had miraculously been averted, and the southern states had agreed to end the slave system peacefully, Nantucket's history would have been quite different. The Friends' meetings might have evolved and developed in amity with the Unitarian church, and the town's social collapse might have been avoided. The whale fishery would have subsided more gradually, as it did in New Bedford, and in all likelihood the economy could have weathered the crash of 1857. If history had gone that way, the town would look very different architecturally than it does today. As it is, time stopped for Nantucket on the eve of the Civil War, and almost nothing was built during the next century. The stark poverty of those years was an architectural preserver, maintaining the mid-nineteenth-century look of the town far better than any zoning board could have done. And in the mid-twentieth century, the quaint, old-fashioned feel of Nantucket became its greatest asset as it changed from a self-sufficient maritime town to a booming seaside tourist destination.

In the winters of 1859 and 1860, the outlook was bleak. On January 4, 1861, a few days before the first states seceded from the Union, Mitchell wrote to a friend,

> If the holidays are sad to you, they are no less so to me. My mother is increasingly feeble and the dread of loss is as great a source of suffering as loss itself. I have abandoned my pleasant sitting room, and spend my

time in her room, studying in a corner, when I am enough easy in my mind to touch a book at all—figuring when it is a duty to figure. There is the painful peculiarity in her illness, of occasional wandering of the mind, and I remember with great anguish the gradual decline of your Aunt Anne. My mother will scarcely live long enough for that. We try to be cheerful and to keep up her spirits and I go out sometimes in the evenings, but we do not leave her alone a moment.

What a sad period it is for the whole country! I really never thought that these things could get so far. We may as well meet the trouble now as postpone it for a few years. I hope we may not have civil war...

Our winter is duller than usual. If there is any society I am wholly out of it, so much so that as I look around on the audience at a concert, it is almost an assemblage of strangers...Can it be that we are all old?[9]

As this letter shows, the decline of Nantucket, the approach of the Civil War, and the death of her mother were all tied together for Mitchell. She had no doubts about the justice of the abolitionists' cause; she had always objected to slavery and opposed racism. But as she waited for the war to start, the issues seemed more complicated and sorrowful than they had seemed in the 1840s when the abolitionist movement had felt joyful, even glorious.

The fight against slavery had been part of Mitchell's life since her girlhood. She was only four years old on the dark morning before sunrise in 1822 when her family was roused by an African American neighbor named George Washington as he pleaded for help to hide the escaped slave Arthur Cooper, who had been living on the island for two years. Maria's father quickly organized a group of Quakers and then went to confront the armed posse of bounty hunters and sheriffs at the front of Cooper's house. While he talked to them, his friend Oliver Gardner sneaked through the back garden and quietly helped Cooper and his wife and children to escape quietly. The *Nantucket Inquirer* reported that the Coopers "had escaped into the swamps, where, it is supposed, they remain concealed among the vast subterranean vaults which have been made by the peat diggers."[10] But Mitchell family lore held that the Coopers were actually hidden at the Gardners' house, and then moved carefully from attic to attic. The Mitchells were quietly proud that they had helped protect their neighbor from being re-enslaved as early as 1822, before abolition had become a national issue.

When Mitchell left her little school to become the librarian of the Nantucket Atheneum in 1836, she joined her fortunes to its founders, Charles C. Coffin and David Joy, both of whom were committed reformers who strongly opposed slavery. The Atheneum itself was not explicitly an anti-racist organization; its charter and by laws never mentioned race until January 1841, when there was a rumor in the *Nantucket Inquirer* that there might be "a resolution to be introduced, the direct tendency of which is to foreclose all the privileges of that house to persons of color."[11] Perhaps the resolution was introduced—many records from before 1846 are lost—but if so, it must have been defeated, for in July of that year, a small notice in the same newspaper announced that there would be a meeting of the Atheneum's proprietors "to ascertain whether the proprietors will instruct the trustees to let the hall to the Nantucket Anti-Slavery Society. Garrison and others are expected from abroad."[12] David Joy in particular was one of the most notable reformers on Nantucket, and he was probably the one who issued the offer to hold the Anti-Slavery Society's great annual convention at the Atheneum in 1846.

The meeting's chief organizer was Anna Gardner, a friend of Maria Mitchell's and the daughter of Oliver Gardner, the man who had helped the Cooper family while William stalled the bounty hunter. The experience of sheltering the Coopers had made Anna into a lifelong abolitionist. Maria was never as much of an anti-racism activist as her friend, but she believed that her work for women's rights was linked to anti-racist work. Since Mitchell was the librarian at the Atheneum at that time, she became one of the hosts of the convention once the board overcame its hesitation. As expected, the convention was controversial. When the mixed-race crowd of abolitionists attempted to board the ferry from New Bedford together, rather than in racially segregated groups, the captain of the steamer refused to allow black passengers onto the upper decks. The conventioneers held an impromptu nonviolent demonstration, and eventually everyone was allowed aboard together. Once they arrived, the great highlight of the meeting was the introduction of the recently escaped slave Frederick Douglass, who had never delivered a public speech to an audience that included white listeners. He was eloquent, and William Lloyd Garrison made the most of his triumph, hiring him on the spot.

Because of the antislavery meetings at the Atheneum, Mitchell had the chance to meet Douglass and Garrison, along with Lydia Maria Child and Wendell Phillips. Another important figure who attended the Nantucket Anti-Slavery Convention was Sojourner Truth. Truth may have been the most fascinating for Mitchell, because, along with Mott and Gardner, she linked the fight against racism to the fight against sexism.

Sojourner Truth is well known for her "Aren't I a Woman?" speech. Though there are many different versions of that speech, the central question that Truth asked about the relationship between gender and race was one that troubled nineteenth-century thinkers a great deal. The prevailing gender ideology of the time defined women as delicate creatures, while the prevailing ideas about race defined black folks as rough, strong, and animal-like. In the bodies and lives of black women, these two ideologies clashed and crumbled. In later years, Mitchell would ask questions very similar to Truth's in her "Aren't I a Woman?" speech, because as a professional woman she was also a somewhat contradictory figure.

It was hard to know how to respond to the war, emotionally or practically. One of Mitchell's brothers joined the Union Navy while another worked as a nonviolent missionary to the battlefield. Other siblings wanted nothing to do with the war. Mitchell was a woman at home during the first modern war that demanded imaginative participation from the home front. She reluctantly supported the Union cause, in part by agreeing to continue her computations for the U.S. *Nautical Almanac,* which was of great use to the navy. But she longed for some other arena for her energies.

As she watched and waited through the long spring of 1861, Mitchell kept replaying her early contacts with abolitionists, particularly Frederick Douglass and Sojourner Truth. She wondered if she could construe her efforts to be a successful professional astronomer as part of a general push for expansion of rights to all castes and classes of people. She longed to do good in the world, but she wondered if astronomy was, or ever could be, part of a social or political movement. Was she destined to be left out of the political sphere after all?

On July 7, 1861, Lydia Mitchell died. After her death, there was no reason for the Mitchells to stay on the island. Maria bought a house in Lynn, Massachusetts, near her sister Anne, and she and her father moved off-island for good. She dismantled the little observatory behind the Coffin School and built another along similar lines to house her Clark telescope. The house and the observatory both belonged to Maria, but her father shared them with her. Sometime after her discovery of the comet, their astronomical collaboration had shifted: he started to assist her, and she led their inquiries. With the move to Lynn, she became the provider for both of them, while her father relaxed into a companionable retirement. Maria and her father had always enjoyed each other's company, and the small house in Lynn was a happy one, though the sorrows of Lydia's death and the great war shadowed everything. By 1863, when Maria published her first long, original scientific essay in *Silliman's,* using observations from both Nantucket and Lynn, life seemed fairly stable for the Mitchells.

In 1864, Augusta Evans published a Confederate novel, *Macaria,* that argued in favor of slavery but against curtailing women's rights. Evans's heroine, Irene Huntingdon, must have been a very troubling figure for Mitchell: not only was she an astronomer, but she was an astronomer who explicitly modeled herself on Maria Mitchell. In Evans's novel, a noble scientific woman was not a threat to any established order or hierarchy; on the contrary, Evans imagined science as an enemy of reform and used the figure of the astronomical woman to argue that an intelligent woman could use her gifts to help maintain social hierarchies. Evans opposed forcing white women to marry or barring them from education, but she did not see these positions as socially radical.

In *Macaria,* Evans wrote, "In glorious attestation of the truth of female capacity to grapple with some of the most recondite problems of science stand the names of Caroline Herschel, Mary Somerville, Maria Mitchell, Emma Willard, Mrs. Phelps, and...Madame Lepaute."[13] The heroine, Irene, who follows in all of their footsteps, uses her astronomical knowledge to support racial hierarchy:

> That the myriad members of the shining archipelago were peopled with orders of intelligent beings, differing from our race even as the planets differ in magnitude and physical structure, she entertained not a doubt, and as feeble fancy struggled to grasp and comprehend the ultimate des-

tiny of the countless hosts of immortal creatures, to which the earthly races, with their distinct and unalterable types, stood but as one small family circle amid clustering worlds, her wearied brain and human heart bowed humbly, reverently, worshippingly, before the God of Revelation.[14]

In this passage, Irene's keen intelligence and her astronomical knowledge work together to ensure her reverence for the well-organized universe in which different "orders of intelligent beings" are arranged into "earthly races" of "distinct and unalterable types." Being a woman astronomer helps Irene to be a conscientious Christian and a brave supporter, even worshipper, of the Confederate slave system.

Macaria was a runaway best seller, both in the Confederacy and in the North. Mitchell was used to being mentioned in print, but it must have come as a shock to be mentioned as a role model for the heroine of the most popular proslavery novel of the war. Mourning for her mother and her desolate island home, Mitchell was increasingly worried about the fate of the country, particularly about her role in the great upheavals. She continued to work steadily, observing double stars, collecting data, and making her first forays into public lecturing in Lynn.

Mitchell longed for a way to make her role as a woman astronomer into an emancipatory project. She inquired about a teaching job at the Troy Female Seminary, but her friend George Bond discouraged her, saying the place was too religious for her. When a man named Rufus Babcock asked if he could visit her to talk about astronomy, she was slow to realize that he was considering her for a professorship at the nascent Vassar College.

CHAPTER 10

Vassar Female College

Matthew Vassar was a wealthy brewer with few ties to higher education. Starting in the 1840s, he spent twenty years searching for a project worthy of his philanthropic ambitions. Touring London in 1845, he noticed a magnificent hospital building, Guy's Hospital, that he believed had been built and endowed by a distant cousin.[1] Inspired, he paid three hundred pounds for copies of the architectural plans and returned to his home in Poughkeepsie, New York, hoping to establish a similarly grand hospital there. But Poughkeepsie was a small town and didn't produce enough sick people for a big hospital. Vassar searched and schemed and canvassed his neighbors and acquaintances, looking for a good reason to build a monumental institution. In 1855 he met Milo P. Jewett, who persuaded him that instead of a hospital, he should build a women's college. Jewett argued that there was "not an endowed college for women in the world," and that if Vassar were to endow one, he could establish the first women's college that could be compared to Harvard and Yale.[2]

Vassar bit. He had never been particularly interested in education or in women's rights, but he loved the idea of doing good with his money, and he was excited by the prospect of being a very big fish in a very small pond. In the 1850s there were a few colleges that admitted women alongside men, and a few women's seminaries that offered higher education to women alone, but Jewett's and Vassar's institution would be the largest and richest women's college in existence.

Milo Jewett was committed to women's education. In 1839 he had founded the Judson Female Institute, in Marion, Alabama, a Baptist seminary based closely on Mount Holyoke. But by 1855, Jewett's antislavery convictions had forced him north from Alabama to Poughkeepsie, where he took over the Cottage Hill Seminary, a small school for girls that Vassar's niece, Lydia Booth had originally directed.

Vassar asked his architect James Renwick to build something that would recall the Tuileries or the Hôtel de Ville in Paris. Renwick, the architect of the Smithsonian, obliged with a palatial Second Empire–style structure containing classrooms, laboratories, a library, and an art gallery, as well as enough room to house the first class of 353 students and their professors. Vassar supervised every detail of the planning and construction, insisting on long, wide corridors running the length of every floor so that students could exercise in inclement weather by hiking from one end of the building to the other.

Vassar was particularly pleased that Jewett's idea gave him scope for his architectural ambition: he had always wanted to build something grand. The sheer size of the new building endowed Vassar College with much of its importance. In *Godey's Lady's Book,* one of the most broadly circulated magazines of the day, the editor Sarah Josepha Hale wrote, "We trust, and we seriously believe, that this enterprise of Mr. Vassar is the initiative of a most important era of improvement in humanity."[3] Matthew Vassar had longed to be a progressive social leader, and Vassar gave him a way to change the world. Inside the back cover of his diary for 1864, Vassar noted down the Bible verse Luke 13:24—"Try to enter in at the strait gate"—and then apostrophized, "The Founder of Vassar College and President Lincoln—Two Noble Emancipists—one of Woman—[the other of] the Negro."[4] Vassar was divesting himself of wealth for the sake of his own salvation, but he also wanted to be a "noble Emancipist" on the scale of Lincoln. He believed that women's education was a cause at least as vital to the national good as the abolition of slavery.

Much to Vassar's delight, the press celebrated his college even before it had opened. Sarah Josepha Hale was his most notable supporter, throwing her editorial might behind Vassar with a series of feature articles. Hale commended the project wholeheartedly, describing it as one of the few "rainbows of hope over the dark clouds of our country's horizon" as the war continued to rage.[5]

Despite the hype in *Godey's,* Vassar was not the first women's college in the United States. Historians argue over which early-nineteenth-century institution of higher learning deserves the distinction, partly because of the difficulty of defining a "college." Some of the most rigorous women's institutions of higher learning called themselves "seminaries," most notably Emma Willard's Troy Female Seminary, which opened in 1821, and Mary Lyon's Mount Holyoke Seminary, which opened in 1837. These early seminaries were somewhere between high schools and colleges. They offered higher education to women, but at first they carefully avoided comparing themselves to men's colleges. Eventually they outgrew their institutional identity—Emma Willard's Troy Female Seminary redefined itself as a high school in 1872, while Mount Holyoke became a college by 1893. But in the years before the Civil War, the distinctions among seminaries, academies, and colleges were hazy. In 1838, a school in Macon, Georgia, opened under the name Georgia Female College. Educational historian Barbara Solomon dismisses it from consideration on the grounds that its curriculum was not particularly rigorous and its students were young: "Despite its name, this pioneer institution resembled a superior academy more than a male college, regularly admitting twelve-year-olds."[6] The women who graduated from Mary Sharp College in Winchester, Virginia, and Elmira Female College in Elmira, New York, in 1855 had stronger claims to be considered the first graduates of women's colleges. Of course, these women were not the first in America to graduate from college: women had been admitted alongside men at Oberlin College starting in 1833 and at Antioch College starting in 1853. The University of Iowa (1855) and Wisconsin (1863) followed suit.[7]

Part of the confusion over what constituted a college was related to the great changes in college curricula in the mid-nineteenth century. Back in colonial times, when Harvard was the only college on the continent, the purpose of a college had been preparation for the ministry. With its heavy concentration on biblical languages (Greek and Hebrew) and its expectation that students read philosophy and physical science in Latin texts, Harvard's curriculum can easily be characterized as classical. According to Stanley Guralnick, a historian of early American science education, "mathematics and science played a small role in Harvard's curriculum" before the nineteenth century.[8] By the middle of the eighteenth century, a small number of other colleges had been established, and "the curriculum of the Colonial College had begun to shape itself along broader lines than those of mere preparation for ministerial offices."[9] The purpose of an eighteenth-century American college was to form young men into gentlemen and prepare them (sometimes secondarily) for some sort of profession, most often law. Natural philosophy entered the curricula of most colleges as a form of metaphysics, but for the most part a college course was focused on classical languages. With the lessening emphasis on preparing students to be clergymen, biblical languages became less common, so the focus was on Greek and Latin.

All of this explains why boys who were preparing for college in the early decades of the nineteenth century tended to neglect science. In male colleges of this time, science was merely an afterthought; according to Stanley Guralnick, "in 1800, not a single college had a teaching staff capable of rendering high-level instruction in science."[10] Girls did not have the option of preparing for college, and therefore they were free to study more science than their male counterparts. As Kim Tolley succinctly puts it, early-nineteenth-century American high schools offered "Science for Ladies, Classics for Gentlemen."[11]

But things were changing. In the early nineteenth century, hundreds of new men's colleges opened across the United States. By 1828, it was hard to define what made a college a college: students ranged in age from ten to thirty, and courses of study were no more standardized than the qualifications for college-level teachers. To add to the confusion, men's colleges had started to rethink their devotion to the classics. In 1828, the *Yale Report* offered a careful rethinking of collegiate education. The authors of the report expressed "willingness to change" and celebrated the introduction of sciences. However, when they asked themselves "What, then, is the appropriate object of a college?" they were unable to reach a consensus. Agreeing that the primary object of a college was "to lay the foundation of a superior education," the *Yale Report* concluded a bit lamely that "as knowledge varies, education should vary with it."[12] Recommending that students study a wide range of subjects, the report insisted that classics ought to continue to be part of the college-level curriculum. As Lynn Peril puts it, "Classics remained the benchmark for the time being, and thus became the criterion of contemporary nineteenth-century higher education by which the women's colleges were judged—and to which the elite schools aspired."[13]

In the midst of the curricular confusion, it made sense that relatively conservative seminaries like Mount Holyoke would distinguish themselves from colleges by offering more ladylike coursework, heavier on math and science than on classical languages. When Mary Lyon, the president of Mount Holyoke, described the school as a "castle of science," she was celebrating its modest femininity, not its innovation.[14] According to historian Miriam Levin, Lyon and the trustees of Mount Holyoke "saw scientific education as producing graduates equipped to order the future, and the study of nature as providing them the harmonious vision of heaven on earth."[15] In this mind-set, science was anything but emancipatory; on the contrary, it helped keep the lower orders in their places. Lyon's "castle of science" was not originally intended to compete with any male college, but twenty years later, as Matthew Vassar was opening his school, the nature of both science and college had drastically changed.

Vassar College was radical because its curriculum emphasized the classical languages much as the curricula of men's colleges did. It was a little different from most men's colleges in that it also offered serious coursework in fine arts and the sciences, which were best understood as courses that feminized the curriculum. And yet by 1865, change was in the air.

One of the best examples of the changing nature of women's relationship to science is the debate that erupted over the naming of Vassar College. When Vassar opened its doors in September 1865, the words "Vassar Female College" were chiseled over the door of the main building. Many of the college's most enthusiastic supporters—and most supporters of women's education in the 1860s—were horrified that Matthew Vassar had given his institution such a name. In 1821, when Emma Willard had opened her Female Seminary, no one had been offended by the word "female," but forty years later it sounded "vulgar, offensive, and Darwinian," in the words of Elizabeth Daniels, longtime historian of Vassar College.[16] Although the founders probably expected some controversy over the college itself, they were surprised by the vehement public outcry against its name. In spite of her support for Vassar, Sarah Josepha Hale spearheaded the call to remove the offensive word "female" from the college's name. Hale wrote a letter to Matthew Vassar: "Female! What female do you mean? Not a female donkey? Must not your reply be 'I mean a female woman'? Then ... why degrade the feminine sex to the level of animals?"[17] Explaining her opposition to the term in the pages of *Godey's Lady's Book,* Hale asked: "Does it seem suitable that the term female, which is not a synonym for woman and never signifies lady, should have a place in the title of this noble institution? The generous founder intended it for young women. The Bible and the Anglo-Saxon language mark, as the best and highest style, VASSAR COLLEGE FOR YOUNG WOMEN."[18]

As Hale's explanations show, the word "female" had animalistic connotations, largely because it was tied to sexual reproduction in a way that "woman" was not. To Hale, "female" sounded clinical and sex-focused; it did not sound ladylike. Moreover, the term was linked to slavery and to Darwinism. During the Civil War era, "female" was often used for slave women as part of a conscious effort to dehumanize them by linking them to animals. Slaves were described as women in conversation, but in print, particularly in proslavery publications, they were usually described as females. As a corollary, it became very important for white women to establish themselves as "ladies" rather than females.[19] Women who owned slaves based their claims to genteel ladyship on their ownership of slaves, while those who opposed slavery based their corresponding claims on the delicate moral sensibilities that made them object to slavery. African American women of the period sometimes hesitated before claiming to be "ladies," but many were outspoken in their refusal to be classed as "females," insisting instead that they were true women, and their womanhood made them sisters of white women.

In 1864 then, just as the war was ending, the term "female" was connected to racial oppression, and the relationship between the fight for racial justice and the fight for gender equality was growing more and more complicated. Although many of the white women who advocated women's rights had been abolitionists before the war, few were comfortable with language that linked white women to the "inferior" races. Sarah Josepha Hale may have feared that women were the next African Americans—that the language of animalization that had been employed to keep dark-skinned people in their place might soon be used against women. Hale was outraged at the idea that women students would be linked to animality rather than to intellect or even to social status. But there were also racist undertones to her objection. Her description of the word "woman" as biblical and "Anglo-Saxon" signals the value that she placed on "Anglo-Saxon" whiteness.

As a matter of fact, whiteness was important to the whole Vassar project. Although Matthew Vassar proudly linked his mission to the emancipation of slaves, the college he founded was decidedly aimed at white women and excluded women of color. This was not the only option at the time. Oberlin College had been proudly interracial since its founding in the 1830s, while Amherst and Bowdoin had started admitting black men in the 1820s. Bates College, which opened in 1855, was both coeducational and interracial.[20] In the years after the Civil War, Fisk (1865), Howard (1867), Atlanta (1867; later called Spelman), and Tougaloo (1869) all opened with the express purpose of offering college-level education to black students.[21] Vassar, on the other hand, excluded black women until 1940.[22]

Although the controversy over the naming of Vassar may have drawn on the racial undertones of the term "female" during the 1860s, the importance of Darwinism to the redefinition of that word cannot be overestimated. When *The Origin of Species* was published, in 1859, it reinvented biology and might even be said to have reinvented science. Around this time, science was in the process of morphing into something dynamic and destabilizing. Theory became far more central than observation. The well-organized scientific systems that Mary Lyon thought would help her girls make the world simpler and more orderly were shattered; scientific thinking changed as the fundamental nature of science shifted from a strategy for describing and cataloging nature to a highly theorized method of inquiry. Of course, Darwin's work was not the only catalyst for the great changes in science, but almost as soon as he published his seminal text, it became the locus of controversies about humans' relationship to the animal world that have raged ever since.

Darwin uses the word "female" incessantly in *The Origin of*

Species. The book does not explicitly discuss human females, but astute readers were quick to understand that the theory of natural selection backed up an evolutionary model that linked humans to animals. Since the link is understood in terms of sexual reproduction, the female pigeons in *The Origin of Species* were—astonishingly—related to the female readers of *Godey's Lady's Book,* who before 1859 had found little in scientific works to shock their ladylike sensibilities.

In 1865, "female" was a particularly loaded term, linking humans to animals and making them the subjects of scientific inquiry and investigation. Since the term "female" was tied to the discourses of both slavery and Darwinian evolution, it had distinctly oppressive implications for many women. *Origin of Species* is not particularly sexist, but the scientific investigations of sex that followed it in the latter part of the nineteenth century were extremely oppressive. Charles Darwin himself eventually traced out some of the sexist implications of his theories in *The Descent of Man, and Selection in Relation to Sex* (1871).

At mid-century, science was changing from a leisurely avocation entirely suitable for girls to a dynamic engine of industry that was taking on a new role as an enforcer of human differences.

Not coincidentally, the changes wrought by the Civil War were part of the changes in science: after the war, the United States was an industrial economy. The nation was no longer pegging its future to large-scale slave-based agriculture. The northern industrial vision prevailed, making the applied sciences economically and socially central to the future of the United States.

Around 1860, science changed in two contradictory ways: the actual content of science became far more unstable and dynamic, and at the same time the newly professionalized field of science took on a new role of social authority and guarantor of social stability. Londa Schiebinger has commented that "the professionalization of science—changed women's fortune in science."[23] For many, of course, "professional" meant male: Thomas Henry Huxley, president of the Royal Society, wrote that women were *"ipso facto,* amateurs," and therefore not eligible for membership.[24] Denying women entry into professional societies and associations was symptomatic of the larger shift toward masculine professionalism.

Maria Mitchell had learned firsthand of both these developments during her European trip: first, she had discovered that the push toward the professionalization of science in Europe had turned most male European scientists into rigid social conservatives who were reluctant to make space for women scientists for fear that they might threaten science's claims to professional authority. When she finally met Mary Somerville and was welcomed into wide-ranging scientific conversations, she learned that Somerville believed science itself was in flux. It wasn't just Darwin and evolutionary biology; in physics, chemistry, geology, and even astronomy, theoretical scientific thinking was driving inquiry into new, dynamically changing realms that would have shocked an eighteenth-century natural philosopher.

The paradoxical internal dynamism and social rigidity of nineteenth-century science worked together to make it an enemy of women. And many women responded in kind, beginning to distrust and dislike science and scientific language. The "female" controversy at Vassar is a perfect example of women's growing hostility to language that attempted to define them scientifically. Cynthia Eagle Russett explains that before the mid-nineteenth century questions about women's nature "were largely the province of folklore, theology and philosophy.... The sexual science that arose in the late nineteenth century ... attempted to be far more precise and empirical than anything that had come before." At the same time, the new "sexual science" "spoke with the imperious tone of a discipline newly claiming, and in large measure being granted, decisive authority in matters social as well as strictly scientific."[25]

One of the most pernicious and complicated facets of scientific discourse after 1860 was its insistence on an analogous relationship between women and people of color. Nancy Leys Stepan explains that "'woman' and 'lower races' were analogically and routinely joined in the anthropological, biological, and medical literature of the 1860s and 1870s."[26] Science struggled to maintain a relationship between the inferior sex and the inferior races. Many white women responded with outrage—infuriated by the calumny that their sex was inferior, they often rushed to disassociate themselves from people of color, whose inferiority they accepted as a matter of course. The historian Louise Newman explains that many if not most white

women who were women's rights activists after 1865 worked "to establish the white woman as the primary definer and beneficiary of
woman's rights at a time when the country was growing increasingly
hostile toward attempts to redress the political, social, and economic
injustices to which African Americans were subjected."[27]

All of these complicated relationships and conflicts were written
into the small word "female" that was engraved on the lintel of the
new Vassar Female College—and unceremoniously scrapped within
two years, when the college changed its name to Vassar College. It is
hard to know whether a "Vassar Female College" would have been
a better institution—less fearful of the links between white women
and African American women, more open to promoting science as a
socially progressive field rather than as an enforcer of human differences. Perhaps it would have been a more limited institution, more
insistent on tying women to their animality than to their intellectual
ambitions.

The controversy that raged over the word "female" is a great example of the way that basic concepts were changing in the 1860s. In
the context of these radical shifts in meaning, the word "female" was
so loaded with problematic connotations that it was probably a good
idea to jettison it. The larger questions for Vassar College concerned
the underlying issues: What should a women's college be? What
should women aspire to, and how should they be educated? What
made a course of study a college course? Although there were many
aspects to these questions, one of the most pressing for the founders
of the school was that of the relative importance of science and the
classics.

In one crucial sense, there was no change in attitude from the
1830s to the 1930s: women were always encouraged to study the least
useful subjects, those that were farthest from preparation for professional careers. What changed was the prevailing social opinion on
what was useful and how a professional was defined. As the sciences
became leading professional fields, studying science came to seem
less and less appropriate for women; and as a classical education became less of a requirement for entrance into professions such as law
and public service, it became less utilitarian, less tied to professional
identity, and more appropriate for women. The earlier women's colleges had remarkably strong offerings in science, but that did not indicate their social radicalism. Mount Holyoke (1837), Vassar (1865),

Wellesley (1868), and Smith (1868), all of which offered solid instruction in science, had all opened before 1870. Miriam Levin argues that Mount Holyoke was far ahead of men's colleges in its science curriculum and its hands-on mode of instruction. Under Maria Mitchell's leadership, Vassar would enroll more students in higher mathematics and astronomy than Harvard did from 1865 to 1888.[28]

After Vassar Female College changed its name, none of the subsequent women's colleges specified in their names that they were women's colleges. "Female" had become a term of denigration. One could argue that the discomfort around the word bespoke Victorian prudishness about sex, but the women who objected to it were not particularly prudish. Instead, they were angry at the way their sex was being redefined. Womanhood was being medically pathologized, and women's attitudes toward their biological identity were changing.

CHAPTER II

No Miserable Bluestocking

When Maria Mitchell heard in 1862 about the plan to open Vassar College, she immediately wrote to the fledgling institution to ask if they wanted to hire an astronomy professor. Matthew Vassar was thrilled by her interest, but it was far too early to start hiring instructors, and Mitchell received a quick response to this effect. It would be three years before the school opened its doors; the founders were occupied with designing the building. Although some board members opposed hiring women as professors, others were delighted by the prospect of landing a figure as well known as Mitchell. Everyone on the board was convinced that the world-famous woman astronomer wanted to teach at the new college.

This was the first of many misunderstandings between Maria Mitchell and the administration of Vassar: she had not been inquiring about a job for herself, but rather for her brother-in-law Joshua Kendall, who was stranded in Meadville, Pennsylvania, teaching mathematics and astronomy at a small seminary. It is notable that Mitchell did not imagine herself as a Vassar professor, since she had expressed an interest in teaching when she inquired about positions at Troy Female Seminary. She may have heard that Vassar intended to hire an all-male faculty, or she may simply have been too shy to speak directly for herself. She would come to regret her initial humility.

When Rufus Babcock, a member of Vassar's board of trustees, visited Mitchell's house in Lynn for lunch in August 1862, she thought he was coming to ask for her insights on women's education in general. She wrote afterward that it had never occurred to her that she might be a candidate for a professorship at Vassar or that the lunch she served was actually an important part of a job interview.

Babcock adored Mitchell. In his gushing report to Matthew Vassar, he wrote:

> She is by far the most accomplished astronomer of her sex in the world I have no doubt. And but few of our manly sort are anywhere near her equal—in her loved and chosen pursuits. She has moreover a breadth of culture I was by no means prepared to expect. She has traveled a year in Europe with the best facilities of access to all the learned; and yet with all this refinedness she is as simple-minded and humble as a child. . . .
>
> With all the rest, Miss Maria is not such a poor miserable "bluestocking" as to know nothing else but astronomy. The day I spent with them their domestic was absent—and Maria prepared dinner and presided in all the housewifery of their cosey establishment without parade, and without any apparent deficiencies.
>
> In her astronomical observatory, fitted up at the back end of her garden, she is still more at home, handling her long and well adjusted telescope with masterful ease, accuracy, and success. She furnishes all the astronomical calculations for the nautical Almanac, at the stipulated pay of 500 dollars a year. This she can probably take with her to Po'keepsie, if we shall be so fortunate as to secure her services.
>
> Knowing too, that she is too independent or of too much shrinking modesty to apply for the situation, and especially that we should not expect her to be going about to hunt up recommendations of her friends for the place, I have myself obtained three testimonials as good and high as New England can furnish.[1]

It is unlikely that Mitchell ever saw this letter. If she had, she might have been chagrined to see that Babcock praised her domesticity almost as highly as her astronomy. Would she have been pleased that she had passed the stocking test, and Babcock had declared her "not a poor miserable bluestocking"? Mitchell knew that intellectual women risked ridicule as bluestockings; in her 1860 profile of Mary Somerville, she quoted the editor of the *Edinburgh Review* Lord Francis Jeffrey, who had defended Somerville against the charge of being a bluestocking. When a friend wrote to him that he had heard "ladies of Edinburgh are literary, and that one of them sets up as a blue-stocking and an astronomer," Lord Jeffrey replied that "the lady of whom you speak . . . may wear blue stockings, but her petticoats are so long I have never seen them."[2] In much of her writing, Mitchell argued against the whole idea of the "bluestocking," a pejorative mid-nineteenth-century term for intellectual women that connoted that they were undomesticated and antisocial.[3]

The rise of the term in the 1850s shows that the tide was beginning to turn against thinking women. In the late eighteenth century, "bluestocking" had originally been a playful term used by the women who attended Mrs. Vesey's literary salon in London. None of them wore blue stockings, but Mrs. Vesey had once invited Benjamin Stillingfleet to attend her salon without formal evening clothes. When he came in blue knit stockings, his outlandish garb became a rallying point for the women who preferred intelligent conversation to fashionable attire. For some years thereafter, the term was a positive allusion to the salons led by women like Mrs. Vesey and Elizabeth Montagu. Within fifty or sixty years, however, the implications of being a bluestocking had shifted, and the term had become a harsh epithet. In 1851, the *Encyclopedia Americana* defined a bluestocking as "a pedantic female" who sacrificed "the characteristic excellences of her sex to learning," and in 1857 a Baptist preacher argued that bluestockings "unsex and degrade themselves, by their boisterous assumption of man's prerogatives and responsibilities."[4]

The change in the meaning of "bluestocking" from a playful self-designation to a term of harsh opprobrium reflects changing attitudes toward women's education and intellect. These changes were probably too complicated to chart; in the late eighteenth century, for example, the heyday of the salons, few women had any educational opportunities at all. In the following decades there was a remarkable surge in the variety and scope of intellectual possibility for women. The condemnations of the "poor miserable bluestocking" unsexed and degraded by her devotion to study were part of a backlash that is properly understood in the context of the concrete advances in women's education in the middle of the nineteenth century.

Even so, it is notable that when Babcock recommended Mitchell for a professorship, he praised her for doing all the jobs of the domestic servant and commended her for being "simple-minded and humble as a child"—qualities not usually prized in college professors. Finally, Babcock lauded Mitchell for her shrinking modesty, and he completed her job application without consulting her. Adding injury to insult, he made a careful note of her *Nautical Almanac* salary, reasoning that if she were to come to Vassar, she could keep that job and support herself on her almanac wages combined with a pittance from the college.

In October 1862, two months after his visit, Babcock confessed to Mitchell that he had secured references behind her back and proposed her for a professorship. Her response expressed great surprise:

> Not a word of what you wrote me had reached me before. I supposed that the whole plan of the college was so incomplete that no further steps had been taken. I can but be gratified that you should think me fitted to fill so responsible a position and that the gentlemen you name, whose good opinion I so much value, should be willing to speak well of me . . . if the subject is pursued, and the thoughts of Mr. Vassar and Com. still turn towards me for the Astronomical professorship, I should hope to see you again.
>
> In the meantime, I should be glad to have you keep me informed (so far as you think proper) of the progress of the arrangements in all departments.[5]

Is there a hint of asperity in Mitchell's request to be kept informed? How far can we trust her expressions of surprise and humility? In our less modest age, there is always a temptation to read professions of modesty as false, and it is clear that Mitchell's humility was one of her strongest assets in furthering her career. But just because her peers valued this trait, we cannot assume that it was a sham; it is likely that Mitchell meant what she said, and that she really was too humble to imagine herself as a college professor.

Mitchell, after all, had never attended college herself, and the very idea of women professors was a new and strange one. Milo Jewett, who became the first president of Vassar, had publicly announced that he opposed women professors. Although his opposition seems outrageous today, Jewett and many other members of the board of Vassar believed that Vassar's claim to being a college that could compete with male colleges must rest in part on a qualified male faculty. If Vassar appointed women without college degrees as professors, how could the school claim to be a college rather than an academy or a seminary?

This was not a frivolous debate, nor was it a short one: from 1861 through 1864, Vassar's board argued over whether or not to appoint women professors. If Maria Mitchell had not sent her fortuitously timed inquiry, even though it was on her brother-in-law's behalf, it is unlikely that Vassar would have appointed a woman to any professorship. Jewett's original plan was to hire men as professors and women as teachers (with secondary classroom responsibilities but primary supervision of students' domestic lives), but this proposal received substantial adverse publicity. The controversy continued to rage in 1864, two years after Vassar had told Mitchell that she was being considered for a professorship. Once again, *Godey's Lady's Book* took the lead in publicizing the debate with an article by Horatio Hale, Sarah Josepha Hale's son.

> In the whole scheme of the organization we remark only one defect; but that is of such a serious nature, that we hope to see it amended before the plan is finally adopted. It would seem that not only the President, but all the Professors are to be men. The only women for whom offices are proposed are some "assistant teachers" who are to "give instruction in the junior classes, under the supervision of the Professors," and a Nurse, a Housekeeper, and Matron. It is impossible to understand the ground of this disparaging exclusion. It will hardly be affirmed that women cannot be found possessing the capacity for government, or the intellectual acquirements needed for the highest position in such an institution.

Although Hale was a strong advocate for women professors, he was careful to distance himself from other women's rights causes:

> We have no desire that women should occupy political offices or should be professors in colleges for young men. But it is peculiarly proper that woman should be the teacher and guardian of her own sex . . . Moreover, it is but right that, as professorships in young men's colleges are held as the acknowledged incentives and rewards of men of talent, who devote themselves to the laborious and ill-paid pursuits of science and litera-

ture, so the same positions, in colleges for young women should be regarded as prizes due to ladies who, by their talents and success in the
same pursuits, shed honor upon their country and their sex.

. . . Surely, the President and the Trustees of this College, which is
designed by the generous founder for the elevation of women will not
commence by degrading her. They will not announce to the world that,
owing to some peculiar defect in the character or intellect of woman (a
defect now for the first time discovered) they have not been able to find
a lady in the United States qualified to instruct her own sex in the
higher branches of science and learning.[6]

In the question of hiring women, as in the question of "Vassar
Female College," *Godey's* eventually prevailed. The major casualty of
the battle was Milo Jewett, who wrote an intemperate letter to one
of the members of the board on the subject which disparaged
Matthew Vassar and eventually led to his being asked to resign even
before the college opened. Fortunately for Mitchell, for Vassar, and
for the English language, the college rejected *Godey's* final sugges-
tion: that women professors be referred to as "professoresses." In May
1865, Sarah Josepha Hale had argued that "no lady in England or
America has held this important office. But in Vassar College ladies
are eligible to this dignity. We hope they will not usurp the man's
title, but following the usages of our language, assume their own
womanly style: Professoress."[7]

Hale was mistaken—there were already a few women professors
in the United States, most notably Rebecca Pennell Dean, who had
been teaching science at Antioch College since 1853[8]—but her argument prevailed, and shortly thereafter she published Matthew Vassar's own statement in favor of hiring women. Vassar encouraged the
college trustees, "Inaugurate woman's elevation and power, genius
and taste, at the same moment that you open these doors to her sex.
Give her a present confidence, and not push her back again upon
some future hope. Let the foremost women of our land be among
the most advanced and honored pilots and guardians of coming
women, and I shall cheerfully leave my name to be associated with
the result." Vassar concluded by listing the fields that he thought
women were particularly suited to teach: "Music, painting, languages, literature, the natural sciences and hygiene are her native
elements, and she has not failed to reach the highest points in as-

tronomy and mathematics."[9] Of course, it was no accident that Vassar thought women were best at astronomy and mathematics: he had already lined up a celebrated woman to teach them. But he was also factually correct: the women who had reached the highest levels of educational distinction around the world during the eighteenth and nineteenth centuries had done so in the fields of mathematics and the natural sciences, particularly astronomy. Eventually, Vassar College would develop a curriculum that was comparable to Harvard's and Yale's in classics but stronger than either in fine arts and in natural sciences, reflecting Matthew Vassar's belief that these areas were women's "native elements."

In January 1864, Vassar wrote to Hale,

> The subject of Women Professors & Teachers is now fairly before out Trustees, who at their meeting 23rd proximo will report their views, and decide if it can be safely adopted in our College at the opening. The only question that can possibly arise, is whether we can obtain prominent distinguished Ladies instructors to fill the several chairs. Miss Maria Mitchell of Nantucket had been named by a gentleman of our board of Trustees as Professor of Astronomy, but the planning and erecting of the Observatory having been under the Superintendence of Professor Farrar late of Elmira Female College, N.Y., it is thought that the chair *may be offered* to him.[10]

At their meeting the following month, the board of trustees decided to propose that both Mitchell and Charles Farrar be appointed, Mitchell in astronomy, Farrar in natural science. An odd side effect of the board's reluctance to hire a woman professor was that they ended up hiring two professors in similar subject areas, thereby increasing their commitment to the sciences. The prolonged and awkward negotiations make it clear that Vassar College did not have a clearly defined hiring strategy.

Babcock wrote to Mitchell asking if she would be interested in such an affiliation, and judging from her response, his letter seems to have outlined a few possible types of appointment. Mitchell's response lurks in the archives as a smoking gun in the history of women in higher education. She rejected the idea of working part-time, writing, "I do not fancy half-way work, and would rather make myself useful in an institution than to be a mere outside appendage." But then Mitchell made a huge negotiating mistake, an early instance of what would become one of the largest problems of women in the academy: she protested that Vassar was offering her too much money. "I had not thought of so large a sum (1500) as you say the President supposes I shall require. I do not believe I am worth it!"

Immediately after this fateful declaration of humility and disinterestedness, Mitchell qualified her refusal to accept the salary indirectly.

> But you will please consider that I shall give to the College probably the last of my working years. I feel that I ought not the throw upon relatives, however willing, the support of my old age.... Would it be wise in me to accept a difficult and responsible position, and have the additional "worry" of a painful economy, in view of the future? Besides, would it be right in me, to lower the standard of pay for other women, by accepting a small salary? On the other hand, I feel that my want of practical experience as a Teacher, should have its proper weight, in the consideration of the value of my services. I must be a learner as well as Teacher.[11]

Mitchell's letter reveals deep discomfort with the discussion of her own salary. She belittles herself, saying that she wants to work full-time but she is not worth the offered pay—in effect, she asks to be paid a part-time salary for full-time work. Even as she is making this terrible bargain, she seems to recognize that she is endangering her own financial future. She begs for mercy, confessing her worry that if she accepts a lower salary, she will be dependent on family charity in her old age. She is also aware that accepting a lower salary will harm other women as they follow in her footsteps and enter professorships. Nonetheless, she cannot bring herself to accept the fifteen-hundred-dollar salary—she simply does not believe that she is qualified.

We do not know the full context of Mitchell's letter. Babcock's letter may have contained a direct or implicit request that she consider a lower salary than Matthew Vassar contemplated. Mitchell may have received the impression that if she were inflexible, the college would rescind its offer altogether. But even if such extenuating circumstances existed, Mitchell would later feel she had done herself and her sex a great disservice. On March 4, 1865, almost a year after these negotiations, the newly appointed president of the college, John H. Raymond, finally responded with a formal offer:

> My dear Miss Mitchell,
>   Would you accept an appointment as "Professor of Astronomy and Superintendent of the Observatory" in Vassar College at a salary of $800 per annum, with board for yourself & your father, and some hope of an advance after the first year, if the finances of the College justify it?[12]

Mitchell had succeeded in securing the job offer—but she had also negotiated her salary down to $800 from an initial offer of $1,500. She accepted the job, and on May 18, 1865, the minutes of the board noted, "The salaries of professors fixed at $2000—they to pay rent, fuel and lights. Miss Mitchell $800 with rooms and board for herself and her father." Mitchell was chagrined when she learned that even the initial offer of $1,500 had been lower than the salary fixed for male professors. Rather than improving as the fortunes of the college improved, the disparity quickly grew worse: by 1867 male professorial salaries at Vassar had increased to $2,500, while Mitchell's had stayed constant at $800.[13] She was not charged for room and board, while her male colleagues, who also lived in the college, were charged $400 to $450, but the difference was still infuriating. After deducting their room and board, male professors earned first $750 to $800 more, then later $1,250 to $1,300, more than female professors—at least twice as much. In the late 1860s, Mitchell went so far as to offer her resignation in protest against her low pay. The college made some compromises, but Mitchell was always paid less than her male counterparts, and it made her bitterly angry. She was not shy about her frustrations: In January 1873, the *Boston Globe* reported that Mitchell had recently delivered a lecture on Caroline Herschel, which concluded by comparing the mad King George to her own employers. "When [the king] heard of Caroline Herschel's labors, he gave her a position in the Royal Observatory, and when he found that she was doing a man's labors, he gave her a woman's half pay. Was he not up to our present stage of progress?"[14]

It would be unkind to blame the victim here, but it is notable that Mitchell helped to create this situation with her professions of humility. Even though she was aware from the start that such a move might eventually impoverish her and would also hurt other women who might receive similar offers, she could not refrain from expressing her modesty and her willingness to accept lower wages. In this respect, Mitchell was no different from most of the workingwomen who have come after her. According to Linda Babcock and Sara Laschever, a similar tendency prevails today, more than 150 years after Maria Mitchell made her great mistake. They argue that in 2003, "the impact of neglecting to negotiate...is so substantial and difficult to overcome that some researchers who study the persistence of the wage gap between men and women speculate that much of the disparity can be traced to differences in entering salaries rather than differences in raises."[15] The American Association of University Professors reports that the same holds true for professors in particular: "Difference in initial base salary is a big contributor to the earnings gap between men and women in academia. This difference stems partly from the fact that many men negotiate more aggressively than many women do. Moreover, disciplines that are overwhelmingly male—science and engineering, for example—are compensated at higher rates than those that include many academic women—such as education and the humanities."[16]

Although the salary inequities for women in the professoriate have remained significant since 1865, it is notable that when Mitchell was hired, science was overwhelmingly female, and science professors earned the lowest salaries. Today the situation is reversed, and the humanities, which are dominated by women, are generally the lowest-paid academic disciplines. What is consistent is that female-dominated fields tend to be poorly remunerated. Low salaries in those days reflected the fact that science was just beginning to professionalize—many science professors lacked formal qualifications—and the fact that the subject itself was so new to the curriculum. But Mitchell's salary was determined mainly by her sex. Charles Farrar, who was offered a less prestigious appointment as professor of natural science (because he was less well known and less qualified to direct the observatory), nonetheless received the higher salary: initially $2,000, rapidly rising to $2,500, while Mitchell's stayed constant at $800.

The story of Mitchell's hiring at Vassar offers a sort of creation narrative for women in academia. As we consider all of the circumstances—from Milo Jewett's worry about the legitimacy of women professors and Mitchell's own anxiety on the same score to Matthew Vassar's certainty that hiring a well-known woman as a professor would generate enough good publicity to guarantee the success of the entire enterprise—we gain great insight into the conditions that shaped the place of all women in academia.

The decades after the Civil War were the height of the cult of true womanhood, when the Victorian ideals of female domesticity defined women's relationships to their homes very differently from men's relationships to theirs. Men's relationships to their living spaces were often understood in terms of money, property, or ownership: men rented or owned their homes. Although many single women also rented or owned houses, the idea of a woman owning her home was almost unthinkable—homes contained their women. For Matthew Vassar, it was unimaginable for a woman to provide a home for her family, rather than the other way around. Vassar did try to imagine it, in part because he planned to hire Maria Mitchell, whose father was dependent on her. Still, when it came down to details, his imagination failed. A letter about buying furniture for professors' dwellings illustrates Matthew Vassar's confusion. He wondered "whether the College had better furnish those apartments or the Professors, some of them (if we have men) may have furnitures & family, and if Widows Ladies Teachers some may be thus situated also they may not all be single Women. This is a question of some importance."[17] Vassar was a very successful businessman, and most of his letters are crystal-clear, but this one is garbled with panic. Vassar assumed that if the professors were men, some might have furniture and families—and then realized that "Widows Ladies Teachers"— whatever the alternatives to male professors actually were—might also have furniture and family. At this point he had already entered into negotiations with Mitchell, who owned her fully furnished house in Lynn and supported her father—but she was a single woman. Vassar was thoroughly stymied by the conflict between his customary domestic ideology and the nuts and bolts of trying to build, and perhaps even furnish, housing for unmarried adult women.

Helen Horowitz, a historian who teaches at Smith College, has written insightfully about the paradoxical architecture of Vassar College. James Renwick, architect of the main campus building, based his plans on similar ones he'd recently drawn up for New York's Charity Hospital. Renwick was a prominent institutional architect, but the institutions he had designed before Vassar were hospitals or asylums, and he had a hard time imagining how students might differ from patients or inmates. "In his scheme for Vassar," Horowitz writes, "Renwick looked to no men's colleges for precedents. Rather he returned to his plan for Charity Hospital, and confirmed the association, reaching back to Mount Holyoke, of women's higher education with the asylum."[18] One oversight that raised much public comment was that none of the rooms had closets—it had not occurred to Vassar or to Renwick that the students would own more than one dress.

When Mitchell finally arrived at Vassar, she discovered that the college had not provided her with a bedroom or a bed. She was housed in the observatory, which she had said she would prefer to sleeping in an assistant teacher's room on the hall with students. The assistant teachers were responsible for every aspect of their charges' lives—supervising their meals and their carefully scheduled twice-weekly baths—so it is easy to understand Mitchell's preference for the observatory. Still, the arrangements were odd: Mitchell's father was fitted up with a small bedroom in the unfinished observatory basement, but Mitchell herself was provided with only a small cot upstairs, placed along one wall of the room that she used as her study and parlor and where she taught her astronomy students about the use of the sophisticated clock that stood against another wall. For the first ten years of her tenure at Vassar, Maria Mitchell slept on a cot in a public room that functioned as both office and classroom. This was not the norm at Vassar.

When the college opened, Vassar's palatial main campus building had only one small satellite building: the observatory, three hundred yards away, designed by Matthew Vassar and Charles Farrar to compete with those at Harvard and Cincinnati and to hold its own in comparison with the great European observatories. It was equipped with a very large telescope, an equatorially mounted refractor with a 12⅜-inch object glass—the third largest in the United States, according to Henry Albers, the astronomer who collected and published Maria Mitchell's papers.[19] The dome rested on an octagonal tower twenty-seven feet in diameter. Three wings radiated from the octagonal center of the building; each wing was made up of two levels of seminar-size classrooms. In the first years, the lower rooms were unfinished basement rooms, but within a decade or so they were finished and converted for public use. The roof of each wing was flat and surrounded by a wrought-iron fence, basically functioning as a third classroom level. The dome room had a slightly elevated floor, so that from each of the classrooms—the prime vertical room, the transit room, and the clock and chronograph room—a staircase climbed up four and a half feet into the great dome. The dome itself revolved on an iron track and weighed about a ton and a half.[20]

The building was not meant to be lived in. As Mitchell put it, "The Observatory was not built for a dwelling house and its radical defects in that direction cannot be overcome—it was beautifully planned for an Observatory."[21] The building's scale (institutional rather than domestic) and style (elegant and symmetrical, a strange combination of Georgian academic, because of the brick, and Second Empire French, because of the wrought iron and the graceful curves) are important to consider when we try to make sense of the fact that Mitchell lived in the observatory. When the college opened, all of the male professors lived in apartments in the main building with their families. Mitchell's observatory was her own, giving her a great degree of independence and institutional power. By the measure of space allocation, Mitchell was extraordinarily powerful at Vassar: she had more space for teaching and research than any other professor. On a campus with only two buildings, she occupied one all by herself. She taught only advanced students, refused to teach introductory classes, and continued to draw her salary from the *Nautical Almanac*. She became a popular public speaker, so speaking fees augmented her salary considerably. It is likely that she was able to earn about as much as her male peers all in all—but in order to do so, she needed to work three times as hard, all while living without a bedroom.

Mitchell turned to poetry to express her complicated relationship to her living space. One poem began:

> Within the room
> That I call "home"
> The wind may whistle around the Dome.

The poem goes on to speculate about what the wind may be saying—speculations Mitchell dismisses with "to me, it is only 'Blow'"—and then it offers some insight into her late-night thoughts:

> I let the wind whistle and pass
> I shut my eyes to the frost on the pane
> I shut my ears to the creaking vane
> I shut my thoughts from the past and its pain,
> I think of my girls soon women to be
> Who daily bring peace and joy to me
> Who watch the Bear whirl round in his lair,
> Who get up too soon to look at the moon
> Who go somewhat mad on the last Pleiad
> Who seek to try on the sword of Orion
> Who, lifting their hearts to the heavenly blue
> Will do women's work for the good and the true
> And as sisters or daughter or mothers or wives
> Will take the starlight into their lives.[22]

This extraordinary poem shows the great emotional progress Mitchell has made since her early years, when a much earlier poem had contrasted the wind's rattle to a suitor's prattle.[23] She had now found a coterie of young women she loved. She dismissed her own physical suffering, shutting her eyes, ears, and even thoughts against pain; instead, she celebrated her students' enthusiasm for astronomy, mentioning the whirling bear constellation of the Little Dipper, the moon, the seven sisters of the Pleiades, and the great constellation Orion. It's clear that her astronomical metaphors were chosen to be slightly gendered or gender-bending. The moon has long been associated with the female sex, while the Pleiades are seven sisters—to "go somewhat mad on the last Pleiad" is a funny, somewhat collegiate way of expressing a striving to be among the sharp-sighted. Finally, of course, these are the girls who seek to try on the sword of Orion: they are ambitious, and they want to be heroic.

The concluding quatrain boldly reclaims astronomy as "woman's work," reframing it in the terms of Victorian domesticity as "lifting their hearts to the heavenly blue" in the process of becoming good, true sisters, daughters, mothers, or wives. The poem starts out in the strangely undomestic "home" of the observatory, describes the bafflements of insomniac loneliness late at night while the wind rattles the apparatus, and then turns to the girls Mitchell is training as astronomers, projecting much less lonely futures for them—filled with peace and joy, with a passion for astronomy that is fun rather than embattled or lonely, and ultimately a clear social good. It isn't ironing or washing the floor that brings "starlight into their lives"; it is collaborative scientific work.

Mitchell was proud of this poem—she made several copies for friends. It was a public expression of her private thoughts, one that acknowledged her personal loneliness while also making the case that the scientific women who surrounded her formed a true community that filled her with happiness, counterbalancing "the past and its pain."

Vassar was in many ways a home for Mitchell. But it is significant that she hesitated, using quotation marks when calling her clock room "home." Late at night, listening to the wind roar through the dome, Mitchell probably had many bitter thoughts about her domestic discomforts. She may have longed for a more conventional family and home, but she treasured her strong and continuing friendships, and was able to bask in the satisfactions of her own successful career. There must have been moments of intense emptiness, and other times when, as always, privacy seemed the ultimate luxury. In her journal, she wrote that sometimes "the darkness is unbearable," but in those moments she had resources that were unavailable to most—up a few flights of stairs was the observatory, with a powerful telescope. The stars kept her company.

CHAPTER 12

"Good Woman That She Is"

Vassar College finally opened its doors on September 26, 1865. Maria Mitchell and her father, William, having taken possession of the observatory a few weeks before, worked hard in the first few weeks to welcome incoming students, their families, and the curious visitors who thronged the campus. Vassar did not have an admissions office; the college made room for every young woman who showed up, examining her in various subjects and assigning her to classes based on her preparation.

The first few weeks were hectic. Maria spent the first days in the main building, assisting with class assignments, welcoming students, and reassuring their anxious parents. No on knew quite what to expect or how to behave. Her father stayed at the observatory, showing visitors the grand dome and the large refractor. He wrote to his one of his grandsons:

> During the four days since the opening of the College, crowds of people have been into the Observatory, consisting chiefly of the students and their parents and friends who have accompanied them to the college. Aunt Maria was most of the time in the college meeting them as they arrived shewing them the college with the other professors, while I took care of them at the Observatory, shewing them the instruments and describing them. The first day I found I had talked steadily from 8 o'clock in the morning till six at night except the dinner hour. In the evening, my throat was much swollen, but the second day there were fewer, and now there are only twenty or thirty a day.[1]

William was like a kid in a candy store: he had always dreamed of working with instruments as sophisticated as Vassar's, and now, thanks to his daughter, he could not only work with them but live with them and share them with a multitude of bright, eager girls.

William Mitchell loved Vassar as much as Maria did. Perhaps he was even happier there than she was. In his middle years, he'd served on the board of Harvard University and been an overseer of the Harvard Observatory, both great honors for a self-educated man, but neither job was quite the same as living on campus. At Vassar, William Mitchell had the best of college life: without any teaching responsibilities or administrative duties, he was free to spend as much time with students as he wanted, and the students adored him. Every Sunday evening, a few advanced astronomy students would gather in the observatory's parlor—dominated by a bust of Mary Somerville, before which there was often a small vase of fresh flowers—to drink tea and socialize. Perhaps these evenings brought back memories of the crowded parlors of Vestal Street and the Pacific Bank of the Nantucket years, when nine intelligent children and their cheerful friends had run in and out of the house—but in some ways the Vassar Sundays were even better. The girls were all interested in astronomy, after all, and they considered it a great privilege to spend time at the observatory. The atmosphere of warm affection and strenuous thought was a tonic for the older man.

Sunday nights were a great pleasure for Maria, too. Teaching was demanding, as was administrative work. It wore her down. But Sunday nights with her best students restored her. In the observatory and classroom alike, she often repeated to her students, "We are women working together," and on Sunday nights, they played together. Both William and Maria had many resources for making long winter evenings pass sociably: Nantucket winters had been a great training ground for intellectual parlor games. One of the best games was Snap, in which everyone would think of a celebrity—a historical or literary figure—and write the name on a scrap of paper. Each participant would draw a name and then write a rhyming riddle to help the group guess the name. The rhymes were pure doggerel, but it took a ready wit and a bit of skill to concoct them, and the game made everyone laugh. One night during such a game, one of the girls mentioned that it was another student's birthday: Mary A. Scott was twenty-four. Maria instantly challenged the circle of Snap players to compose odes for Scott's birthday. Everyone scribbled furiously.

That night's game ended up being a momentous one in Vassar's history, and in the lives of the women who were playing it. They seemed to know that it would, as someone carefully saved every single one of these birthday odes, which are still preserved in the Vassar archives. One is scrawled on the back of a petition for women's votes; others are on sheets of paper ripped from a variety of notebooks. Mitchell's is the best:

> No general e'er his battles fought
> With more of skill than M. A. Scott
> No general e'er his honors bore
> With less of pride, of good sense, more.

The verse is short and snappy, but it's also felicitous: reading it a century and a half later, we get a sense of Scott's determination, her competence, and her pleasant demeanor. Mitchell had always been a good hand with a rhyme, but suddenly in this four-line sketch, she'd found a new métier—the professorial ode.

Even before she started at Vassar, Mitchell had wrangled with the administration about grading. She simply refused to do it—she didn't believe that minds could be evaluated on a simple scale. She had two arguments against grades. The simplest was "There is no intellectual unit"—minds were not made to be measured, and so Mitchell averred, "I cannot express the intellect in numbers."[2] Mitchell's other argument for giving all of her students a 5 (when 5 was the top grade) was this: "If a girl has faithfully studied her lesson and does not know it, she deserves 5 for her industry. If she has not studied, yet knows it, she deserves 5 for her intellect. If she has neither studied nor knows it, she deserves 5 for her audacity in coming before me!"[3] But this liberal policy didn't mean that she wasn't cognizant of her students' work. She knew each one well—without assigning a numeric value, she learned to understand each young woman's mind and character. When Mitchell wrote her verse about M. A. Scott, she might have suddenly realized that silly poems would be a perfect way to let each student know that she had paid careful attention to her.

In early June, as classes were ending, Maria Mitchell invited her students to the dome of the observatory to celebrate the end of the first year. Roses were blooming all around the building. She set up tables on the porch and under the dome itself and covered each with snowy linen and flowers. She'd worked for weeks to prepare—the rolls were in the shape of crescent moons, each butter pat was formed into the shape of a star, and the strawberries were perfect. Each place card held a little hand-printed image of a planet. The students breakfasted merrily on this simple but delightful fare, and then Miss Mitchell pulled out a wicker basket filled with scrolls of paper carefully tied with ribbon. She asked each student to draw a scroll, open it, and read it aloud: each was a poem praising the virtues of one of the students. The woman who had just been hymned chose the next scroll, reading a poem about one of her classmates—and so on, through both junior and senior classes, about thirty students in all. Mitchell had written an ode for each one.

For Mary King, a young woman whose letter of recommendation had been written by Ralph Waldo Emerson, her neighbor in Concord, Mitchell composed the following:

> That pleasant town by Sudbury's fair water
> Where philosophy reigns and poets sing
> Has sent us the flower of its worthy daughters
> Has sent us our charming Mary B. King
>
> And when she returns to Concord's wise circles
> May she rival its sages and sing like its poets
> And I have no doubt from the record she bears
> That she'll be abundantly able to do it.

Not surprisingly, Mary King treasured the poem for her entire life—Mitchell was comparing her to sages such as Ralph Waldo Emerson, Nathaniel Hawthorne, Henry David Thoreau, and the incomparable Louisa May Alcott. In Miss Mitchell's perceptive eyes, Mary King was their equal. Such expressions of approval, ambition, and high expectations must have meant the world to Mitchell's students.

As the years went by, the dome parties became one of the most beloved of Vassar institutions. During the year, Mitchell was a rather formidable professor. She was very famous, of course, but the level of mathematics she demanded of her students was also quite difficult. Her idea of a college-level science course was much more rigorous than corresponding courses at Harvard and other men's colleges. With her end-of-year celebration, she showed that she had great affection for and insight into her students. The dome parties also showed the students' easy intellectual transition from astronomy to poetry, which were not considered contradictory tastes or aptitudes. Both required insight, precise observation, imagination, and clear, elegant thinking.

In later years, the students also brought poems about Mitchell. The most beloved of these odes was the "good woman" song, which, like the "Battle Hymn of the Republic," was set to the tune of "John Brown's Body." The song celebrated Mitchell's personality rather than her work. According to Mary King Babbitt, when all of the astronomy students stood to sing the song "with a rousing chorus of 'Glory, Glory, Hallelujah,' the effect was very stirring."[4]

> We're singing for the glory of Maria Mitchell's name;
> She lives at Vassar College and you all do know the same.
> She once did spy a comet and she thus was known to fame
> Good woman that she was
>
> She leads us through the mazes of hard astronomy;
> She teaches us nutation and the laws of Kepler three,
> Th' inclination of their orbits and their eccentricity.
> Good woman that she be.
>
> In the cause of woman's suffrage she shineth as a star
> And as President of Congress she is known from near and far,
> For her exe-cu-tive 'bility and for her silver ha'r
> Good woman that she are.
>
> Though as strong as Rocky Mountains, she is gentle as a lamb,
> And in her ways and manners she is peaceable and calm,
> And our mental perturbations she sootheth like a balm,
> Good woman that she am.
>
> Sing her praises, sing her praises, good woman that she were,
> For though Pope says 'tis human, she is hardly known to err;
> And from the path of virtue she never strayeth fur,
> Good woman that she were.
>
> Sing her praises, sing her praises, good woman that she is,
> For to give us joy and welcome her chiefest pleasure 'tis
> Let her name be sung forever til through space her praises whiz,
> Good woman that she is.[5]

It was telling that the students' song was adapted from "John Brown's Body." Originally, this was a rather gruesome abolitionist hymn that Union soldiers sang about "John Brown's body moldering in the grave," which claimed John Brown as the heroic martyr of the antislavery cause. Later, Julia Ward Howe had turned the tune into a sentimental Christian, nationalist anthem that began, "Mine eyes have seen the Glory of the Coming of the Lord." Howe and Mitchell had become fast friends by the 1870s because of their shared interest in women's rights, particularly women's educational rights; so it is fitting that the women of Vassar turned her Civil War song into the Mitchell song that wrapped all of the reformist fervor of mid-century into a song about Mitchell's good character. In these lyrics, Mitchell became the next John Brown—a martyr for the cause—but she was not a dead martyr. She was a living one who was forced to deny and transcend her physical body. She was a representative intellectual woman in the age of reform—the icon of educated, emancipated, mid-nineteenth-century womanhood.

Mitchell liked the song very much. As she put it, "I have always liked that song because it is true. I am not a wise woman, but I am a good one."[6]

Not only was Mitchell's cot in her classroom, but her students were free to visit at any hour of the night in order to make astronomical observations. Instead of private quarters, then, Mitchell had the companionship of her students. Two of Mitchell's poems, clearly much more personal than the dome poetry, explore the love and admiration that she felt for them. One is a short verse written on a tiny piece of paper—perhaps for a parlor game, perhaps in private:

> The lowliest cot is palace hall
> If thou art with me, it is all.
> The palace hall is lowly cot
> Where thou are not,
> Oh M. A. Scott!

Of all the scraps in the Mitchell archive, this is one of the most intriguingly intimate. Is it a love poem? Mitchell did sleep on a cot in a palatial but lonely place. Did Mary Scott ever cuddle next to her on that narrow bed? If she did, what, if anything, did it mean? In the early nineteenth century, women often shared beds, and passionate kisses and caresses, without thinking of themselves as lovers. Intimate friendship was not necessarily sexual friendship, and yet it's hard to understand this verse as anything but a love poem. As a Quaker, Mitchell was comfortable with "thee"s and "thou"s—but "thou" is the intimate form of "you." "If thou art with me, it is all" is a line from a love poem. When the whole verse is about bed, it's hard to avoid the conclusion that Mitchell very much enjoyed her student's company in bed, and missed her when she was not there.

Does this mean that Mitchell and Scott were lesbian lovers? This is probably a poem of the late 1860s or early 1870s, just as notions of lesbian identity were developing. In 1839, Mitchell could conceivably have written lines like this with no awareness of sexual overtones. By 1899, a poem like this would have been a clear declaration of sexual love. But the sexual meaning of a poem written halfway between those eras must remain obscure to us. The middle of the nineteenth century was the height of the era of intimate friendships between American women, many of which involved erotic components that ranged from intense romantic feelings without physical contact to kissing or handholding to genital play. Some intimate friendships were probably not erotic at all. For most of the women who entered into intimate friendships in the mid-nineteenth century, the concept of a lesbian identity would have been quite alien— but this does not mean that the relationships were unimportant to individual women's sense of themselves.

In Mitchell's later years, such gossip swirled around outside of women's colleges, sometimes touching on Mitchell. She resolutely ignored it. When Benson Lossing, her dear friend and best advocate on the Vassar board, asked her about her reaction to a prurient story about her, she responded:

> Lynn Mass, August 1881
>
> My dear Dr. Lossing,
>
>   I will put the newspaper scrap into the hands of one of my sisters and ask her to strike out what is inaccurate. I do not read such things as their misstatements fret me somewhat.
>
>   I have not the most remote idea of the author of the article, and cannot guess how the mixture of truth and falsehood was obtained. I

have seen my family laughing over it. What a pity that no "blighted affections" can come into the story of my life! It would pay a newspaper so much better! Poor Vassar...[7]

This is a brave denial. It is quite likely that such "misstatements" affected her more painfully than she admitted to when she said that they "fret me somewhat." And of course, such a denial, particularly when it's written to one's boss, must be taken with a fairly large grain of salt. Mitchell's affections, blighted or not, will never be clearly limned for the historian. Her strategy with this article was to "put it into the hands of one of my sisters," probably Phebe, who managed Mitchell's papers; and Phebe, at her sister's instructions, was in the habit of "striking out the inaccuracies." After Mitchell's death, Phebe destroyed all evidence that would have troubled her sister, so there is nothing left in the archives (except perhaps the "lowly cot" poem, which stayed at Vassar) to tell us much about Mitchell's love life, if she had one.

But in spite of the fact that Mitchell and her family tried to protect her heart from scrutiny, I don't believe she was a secret lesbian. I think her love for her students is best explained in another undated poem she wrote for four of her students at the end of an academic year. This is clearly a private poem, as it's written on a scrap and was not rewritten in Mitchell's public hand. It avows her love for her students, and even calls them lovers, but it doesn't describe a sexual love:

> Sarah, Mary, Louise and I
> Have come to the crossroads to say Good-Bye
> Bathed in tears and covered with dust
> We say "Good-Bye" because we must
> A circle of lovers and not of peers
> They in their youth and I in my years
> Willing to bear the parting pain
> Hoping we all shall meet again
> For if God is God and Truth is Truth
> We shall meet again, and all in our youth.[8]

Mitchell's description of her students as a "circle of lovers" may sound a bit scandalous to twenty-first-century ears, but that is arguably because there is something wrong with our ears. Sex with students is a bad idea, but love for them is a necessary part of teaching. That Mitchell saw her students and herself joined as "a circle of lovers" was for her a holy vision: she thought they would spend eternity together in heaven. She certainly didn't think of her love for her students as a guilty secret. On the contrary, her love for women and her commitment to women's intellectual culture was becoming the most important thing in her life.

Maria Mitchell was forty-seven when she started teaching at Vassar. Early in her career, the divide between teaching and research had been absolute. Now, she and Harvard's George Bond were part of the first generation of professors who were also research astronomers. It was difficult to negotiate a combination of teaching and research when no one had really done this before, and the great risk was that when scientists started to teach astronomy they would be unable to practice it themselves. The hours were particularly hard for astronomers—to keep to a daily class schedule, show up for administrative meetings bright and early, and stay up observing through the night (in an era when astronomy was primarily observational) meant forgoing sleep altogether. For the first three years, Mitchell tried to do it all—she continued with her research, kept up with her computing for the almanac, and built supportive and challenging communities of the astronomically inclined women in her classes. At the same time, she was also caring for her father and living without a real residence. She worked day and night, and her private writings grew less personal. Her notebooks became a gathering place for lecture notes, research ideas, agendas, itineraries, and random astronomical insights.

Before too long, this pace took its toll. By 1868, after three years with virtually no sleep, she was coming to realize that things must change. She wrote, "I dare not repeat the brain struggle of last year—it is suicidal to attempt."[9] She needed to choose between the three careers she had juggled for three long years; something would have to give. The decision was agonizing. She hated to give up the almanac work because it guaranteed her financial self-sufficiency. No matter where life took her, she could live anywhere as computer of Venus. She hated to give up her ambitious research program because she finally had the facilities she had always dreamed of. And she hated the thought of turning her back on her students—she loved them, and she believed passionately in her responsibility to educate the next generation of women astronomers. If she didn't do it, there was no one else to do it in her place. However, she feared she was ruining her health.

On August 1, 1868, in her house in Lynn, which she had retained when she moved to Vassar, Mitchell wrote a sentence or two about her worries in her small yellow notebook: "If I return to Vassar I am to teach an *advanced* class, if I wreck myself (as Vassar Prof) in the effort. I am to keep my health, 1st to maintain a high order of learning. I believe in this very work split, but it is the work all the same."[10] Although her syntax is slightly cryptic here, clearly Mitchell was struggling to find balance. Running a research observatory and an instructional astronomy department were jobs for two people. Given the intense demands of the job, the outrageous inequity of her salary, and the radical discomfort of her living quarters, it was very tempting to refuse to return to Vassar.

But she did return. On September 22, 1868, back in Poughkeepsie, Mitchell wrote, "I have written today to give up the Nautical Almanac work. I do not feel sure that it will be for the best, but I am sure that I could not hold the almanac and the college, and father is happy here."[11] Like many of the women scientists who would follow her, Mitchell chose familial happiness over scientific work. In a very real sense, giving up the almanac was a step backward for Mitchell. As a computer of Venus, she had been one of the top professional astronomers in the United States. Vassar had received a great deal of positive publicity, particularly in the pages of *Godey's Lady's Book,* but it was still a very new institution, and a professorship at Vassar was of questionable prestige compared to her work at the almanac. Mitchell had brought her fame and her international reputation to Vassar, but Vassar could not necessarily recompense her in kind; it was as though she were deciding to be a schoolmarm rather than a scientist. Mitchell struggled with the feeling that she was giving up her identity as a professional astronomer.

She was now determined to lift up Vassar's astronomy program to a truly advanced level. Doing so would help establish the prestige of the institution and the worth of her job, but her primary motivation was higher—she longed to offer her students the kinds of opportunities and challenges she had had to create for herself on Nantucket. Part of the reason her task was so difficult was that she had no predecessors in men's colleges; mathematical astronomy was only just entering college curricula. Indeed, Mitchell's quest to institute an advanced astronomy course was parallel to (or slightly in advance of) the efforts of many other professors of science at the time. Miriam Levin explains that the women teaching science at Mount Holyoke from 1860 to 1873 worked hard to push their course offerings to a more advanced level because every science professor in an institution of higher learning in America was trying to do so. In many ways, Mount Holyoke, which had been founded in 1837 as a "castle of science," was ahead of the curve: most significantly, the school had long used interactive demonstrations to supplement lectures. Although the demonstrations at Mount Holyoke arose in part from a sexist hesitancy to permit women to lecture (since lecturing was obviously speaking in public), Levin argues that the interactive pedagogies developed at Mount Holyoke were a significant "contribution women made to the foundations of laboratory teaching in higher education."[12]

Mitchell was not impressed with the passive learning and lecturing she encountered when she visited Harvard in 1867. Her old friend and fellow almanac computer Benjamin Pierce lectured without interruption (unlike Cyrus Pierce, the renowned pedagogue who had trained the young Mitchell). Mitchell was underwhelmed by the class: "Pierce had filled the blackboard with formulae and went on developing them. He walked backwards and forward all the time, thinking it out as he went. The students at first all took notes, but gradually they dropped off until perhaps only half continued. When he made simple mistakes, they received it in silence, only one, that one his son, a tutor in the college, remarked that he was wrong. The steps of his lesson were all easy, but of course it was impossible to tell whence he came or whither he was going..."[13] In her own classes, Mitchell avoided lectures, which she dismissed as the teacher showing off rather than trying "to develop her pupils."[14]

Mitchell's more interactive pedagogy was not framed solely in terms of gender—her father's peculiar gentle charm, the experience of assisting Cyrus Pierce, and the pedagogical writings of Bronson Alcott were major influences. When she declared: "[L]et the imagination have some play; a cube may be shown by a model, but let the

drawing upon the blackboard represent the cube, and if possible let
Nature be the blackboard; spread your triangles on land and sky," she
was channeling Bronson Alcott and perhaps even Emerson.[15] Still, it
meant a great deal to her and to her students that they were "women
studying together."[16] The antiauthoritarian, hands-on method was in
part a women's approach: the fact that they were women gave an ad-
ditional meaning to everything that happened in the Vassar observatory.

One night, a few weeks after she had resigned from the *Nautical
Almanac,* Mitchell scrawled a resolution into her notebook: "RE-
SOLVED: In case of my outliving father and being in good health,
to give my efforts to the intellectual culture of women, without regard to salary."[17] It was a momentous resolution for Mitchell. After
three years of struggle, she was deciding that on the most fundamental level, women—and women's intellectual culture—were more
important to her than astronomy. Historian John Lankford reads this
as a somewhat tragic decision. Mitchell changed her plans for the
observatory, embarking on a simpler course of research that allowed
her to involve Vassar students in every aspect of her astronomical
work but also lessened her chances of making the sorts of discoveries
that would win her lasting astronomical fame. Mitchell was very
clear about this choice. In the same small notebook where she had
worried that she might "wreck" herself by teaching advanced astronomy, she outlined the advantages of her two possible research
programs: one "would be likely to interest the girls," while the other
"would in time, gain [her] a real solid reputation."[18]

Lankford is partly right; Mitchell's decision to prioritize "the
girls," as she described them, was a compromise that diminished her
chances of carrying out a more ambitious program of research. But
if what Mitchell wanted to accomplish was work that was "solid and
[would] endure for 100 years,"[19] as she put it, perhaps her choice was
the right one. In the history of American astronomy, Mitchell's pedagogical innovations were extremely important. She was the first
college professor to let her students loose in the observatory and give
them sustained hands-on experience with instruments. She described some of her classes as "half a conversational lecture and half
questions and answers. I allow them great freedom of questions and
they puzzle me daily. They show more mathematical ability than I
had expected, and more originality of thought. I doubt if young men

that age would take as much interest in science. Are there seventeen students in Harvard College who take mathematical astronomy, do you think?"[20]

Mitchell also described her hands-on methodology: "They are instructed in the use of instruments, small telescopes and a transit are put into their hands, which they are encouraged to use freely in their intervals of leisure. They can, if they choose, engage in the routine work of the Observatory, and are allowed some practice with the meridian circle . . . [Second-year] students are allowed some practice with the equatorial telescope."[21] Under Mitchell's guidance, the observatory became something of a scientific playhouse. Students were in and out at all hours, grabbing small Clark telescopes and climbing the creaking stairs to the roof to observe for themselves. Mitchell was always available to help them, but she was thrilled when students figured things out on their own, and equally excited when they came up with questions that stumped her. One of the most telling relics of her years in the classroom is a worn notebook carefully labeled "book of questions" that contains hundreds of unanswered questions written in the hands of hundreds of different students. By encouraging students to formulate their own questions, Mitchell was teaching fundamental lessons about the scientific method.

Mitchell also led the move away from what she called astronomical geography (learning the positions of objects in the sky) and toward mathematical astronomy. Vassar students were the very first wave of American college students to study what would become known as astrophysics, a natural result of a pedagogy that focused on student fieldwork rather than memorization. Cornell, Dartmouth, and eventually Harvard all followed Mitchell's lead in this respect.

Beyond astronomy was the cause that had become even dearer to Mitchell: women's intellectual culture. Mitchell believed that "women, more than men are bound by tradition and authority. What the father, the brother, the doctor and the minister have said has been received undoubtingly. Until women throw off this reverence for authority they will not develop. When they do this, when they come to truth through their investigations, when doubt leads them to discovery, the truth which they get will be theirs, and their minds will work on and on unfettered."[22] Mitchell felt that teaching women to think for themselves and throw themselves into their own work was more important than teaching them the particulars of astronomy. She explained to her students: "I cannot expect to make you astronomers, but I do expect that you will invigorate your minds by the effort at healthy modes of thinking."[23] She told them stories about Mary Somerville and Harriet Hosmer, Rosa Bonheur and Elizabeth Barrett Browning. These "women of genius" were exemplary, and yet, Mitchell explained,

> I do not hold these women up to you as examples in their specialty. I am far from thinking that every woman should be an astronomer or a mathematician or an artist, but I do think that every woman should strive for perfection in everything she undertakes. If it be art, literature or science, let her work be incessant, continuous, life-long. If she be gifted and talented above the average, by just so much is the demand on her for higher labor, by just that amount the pressure of duty increased. Any special capability, and sense of peculiar fitness for a certain line is of itself an inspiration from God, the line is marked out for her by his finger. Who dare turn from that path? . . . Think of the steady effort, the continuous labor of those whom the world calls "geniuses." Believe me, the poet who is "born and not made" works hard for what you consider his birthright. Newton said his whole power lay in "patient thought" and patient thought, patient labor, and firmness of purpose are almost omnipotent.
>
> Are we women using all the rights we have? We have the right to steady and continuous effort after knowledge, after truth. Who denies our right to life-long study? . . . We have another right, which I am afraid we do not use, the right to do our work well, as well as men do theirs. I have thought of this part of the subject a good deal, and I am almost ready to say that women do their work less thoroughly than men. Perhaps from need of right training perhaps because they enter upon occupations only temporarily, they keep school a year, they write one magazine story, they keep accounts for a few months for some uncle, they take hold of some benevolent enterprise for one winter, when it's "all the rage."
>
> The woman who does her work better than ever woman did before helps all woman kind, not only now, but in all the future, she moves the race no matter if it is only a differential movement, it is growth. And this seems to me woman's greatest wrong, the wrong which she does to herself by work loosely done, ill-finished, or not finished at all.[24]

Mitchell followed her own creed, working incessantly and passionately and eventually finding innovative ways to combine teaching and research. Although the program of research she eventually chose for her first years at Vassar was less ambitious than the other she had contemplated, it was still important original work that she carried out with painstaking care and successfully used in the classroom, allowing her advanced students to work with her and to get a taste of collaborative original research.

Mitchell published seventeen scientific papers independently during her time at the college. Some were in general-audience journals, but she also published regularly in *Silliman's Journal,* the top scientific journal of the day. By 1876, the Vassar College Observatory had agreed to furnish the monthly astronomy column for *Scientific American*—under Mitchell's guidance, her students computed the positions of planets and wrote the monthly column. The articles were unsigned, but if we consider Mitchell the lead author on all of them, her publication numbers swell: there were fifty-six. If Maria Mitchell had compiled a curriculum vitae, it would have listed at least eighty publications of various types. Publication was a relatively new phenomenon for college and university faculty in the late nineteenth century; the doctrine of "publish or perish" had not yet taken hold, and there were comparatively few scholarly journals in which to publish. But Vassar trustees were in the vanguard here also, and Vassar was one of the earliest colleges to keep careful records of faculty publications. At first, the college only recorded the publications of its male professors. The Vassar College librarian Frances Wood recalled that Mitchell "judged everything from the standpoint, 'How is this going to affect women?' I remember her indignation at being overlooked with the other two women of the Faculty by President Raymond in his demand for a list of what the members had published. 'We may not have done so much as some of the men, but we have all done something.' It took a great deal of apology on his part to soothe her wounded feeling and restore her natural good humor."[25] A few years of struggles with the Vassar administration over her salary, teaching load, and even—as we will see in the next chapter—her religious beliefs had changed Mitchell's attitude almost completely. When the college administration ignored her, she protested not so much on her own behalf but because she knew that accepting poor treatment would affect other women scholars who followed her.

For the first years after its founding in 1865, Vassar was without question the preeminent women's college in the world. There was nothing like it. Because it was unprecedented in so many ways and answered a pent-up demand, it attracted many of the brightest women in the United States. The class of 1868 was the first class of women to complete the four-year college curriculum, and by all accounts they were a superlative group. Mitchell would later remark that "the college is what it is, mainly because the early students pushed up the course to a collegiate standard."[26]

Among all the intelligent women of the class of 1868, Mitchell's advanced astronomy class was dearest to her heart. There were six women in it, and Mitchell's father dubbed them the Hexagon. The Hexagon practically lived at the observatory, and all of them confided in both Mitchells. One, Mary Whitney, spent the rest of her life at Vassar, eventually succeeding Mitchell as director of the observatory. The others all stayed in close touch with Mitchell through their lives, and when Mitchell described herself and her students as "a circle of lovers, and not of peers," it was the hexagon she had in mind.

On Christmas Day 1867, all six members of the Hexagon spent the day with William Mitchell, who was growing frail. "We had a cheery and jolly time of it," he wrote. "I almost forgot my age and my infirmities."[27] But it was William's last Christmas. That spring he died at the observatory, where he had spent four of his most pleasant years. The *New York Times* reported, "Mr. Mitchell died at the age of 76, having lived, according to his friends, a life of singular beauty."[28] One of the most beautiful aspects of William's life was his collaborative relationship with his daughter. Later, Maria's sister Anne commented:

> As long as he lived William Mitchell continued to instruct his daughter, and Maria, his only unmarried child, was never weary of expressing her gratitude. A few hours before he died, in April 1869, she ventured to ask him a question concerning a matter which had puzzled her for many months, but whose solution she expected the moment her father was sufficiently recovered to give it his attention. He had already told her that he had but a few hours to live, adding, as he finished what he had to say, "but do you go to Iowa, in August, to observe the eclipse all the

same." Encouraged by this manifestation of his undying interest in her studies, she ventured to put her question. He turned away, as she thought, in weakness, and she felt herself rebuked. But no! In a few moments he resumed his original position, and in a clear voice, explained the matter fully. This was at 2 o'clock in the afternoon and he died at sunset.[29]

His last words were to Maria: "Thee'll look for a comet tonight, will thee not, my child?"[30]

In his last years, William Mitchell had shared his fondness for astronomy and his belief that young women could be world-class astronomers—a belief so momentous in Maria's life—not just with his daughter but with all of the students who haunted the Vassar Observatory. With his farewell, William Mitchell encouraged his daughter to keep working, to hope for new discoveries. The girls of the Hexagon, who thought of William as a father figure, were at her side that spring, searching every night for comets, observing Jupiter and Saturn, and making their most ambitious plan yet: after commencement in June, Mitchell and her Hexagon of newly minted college graduates would travel to Iowa to observe the total solar eclipse for the Smithsonian. It was a fitting tribute to William: the experience of counting seconds for her father during the total solar eclipse of 1831 had been a turning point in Maria's life. Now, more than thirty years later, Maria would offer her students the same sort of experience that her father had offered her. They would work together as professional women astronomers.

Although the Iowa expedition was only a few months after her father's death, Mitchell was in good spirits on the trip. She loved the students she was traveling with, and she greatly respected their abilities. In many ways, the expedition reminded her of her glorious summer in Maine in 1849, working for the Coastal Survey as a young woman. But on that trip, she had been one of the only women, and the few others who had come along were wives rather than astronomers. As far as Mitchell knew, an all-female scientific expedition had never been mounted before.

The eclipse expedition of 1869 was the high-water mark of Mitchell's years at Vassar. Her Hexagon of students, all of whom had started at Vassar at its founding, were probably the most talented and tight-knit group she would ever teach. The Smithsonian was glad to have their services, as their work was important and their skills were relatively rare. The Vassar women were at the top of the game. That year, Mitchell herself was inducted into the American Philosophical Society.

This is not to say that the women of the Hexagon were the only astronomers observing the eclipse: the train west had been crowded with astronomers who were congregating from all over the United States. They were also not the only astronomers employed by the United States government—the Nautical Survey had sent a team led by Charles Coffin, with whom the Vassar team would be loosely associated. But the Vassar expedition was completely independent and wholly female.

In her article on the eclipse published in *Hours at Home* in October 1869, Mitchell was playfully coy about the sex of the expedition members. Without identifying herself or her team, she explained the significance of eclipses not only to astronomers but also to chemists, physicists, and photographers. She went on to offer lucid explanations of why eclipses were interesting to chemists, physicists, and photographers as well as astronomers, and then to describe the experience of using cameras, telescopes, and spectroscopes to observe the eclipse. Her narrative is riveting, and her description repeatedly emphasizes the competence, skill, and cool-headed determination of her team. Finally, she closes the article by remarking,

> Piazzi Smyth says "the effect of a total eclipse on the minds of men is so overpowering, that if they have never seen it before they forget their appointed tasks, and *will* look around during the few seconds of obscuration to witness the scene." Other astronomers have said the same. My assistants, a party of young students, would not have turned from the narrow line of observation assigned to them if the earth had quaked beneath them. They would have said
>
> > By the storms of circumstance unshaken
> > And subject neither to eclipse nor wane,
> > *Duty* exists.

In the closing sentence of her essay, Mitchell puts the zinging question: "Was it because they were *women?*"

Readers of *Hours at Home* would likely have been taken aback, as it takes a careful reading to notice that there are no gender-specific pronouns anywhere in the article. Instead of the telltale "she," Mitchell had used a collaborative "we" throughout, playfully concealing the gender of herself and her team. The final twist changes the essay in unexpected, potentially disturbing ways. A reader ignorant of the sex of the group would probably read the article for its astronomical content and be pleased with its lucid explanations, its colorful descriptions of the beauty of the eclipse, and its compelling narrative of cool-headed focus and determination. Once the reader learns that the expedition is made up of women, the article reads quite differently—it becomes an argument for women's scientific skills as well as an account of one event. Mitchell was well aware that readers would be unable to process the scientific information if they knew the scientists' gender. In her penultimate sentences about the "effect of a total eclipse on the minds of men," her argument is suddenly clear: women's minds are tougher than men's. Though men may find it impossible to concentrate during an eclipse, women's focus on their duty cannot be eclipsed. Mitchell's playful question "Was it because they were *women?*" posits that women may actually be *more* reliable as scientific observers than men are. The genius of her article is that she also shows that women scientists' gender shapes the reception of their work.

Mitchell writes about the eclipse with notable merriment: the Vassar women had fun together. No one was rude to them on account of their sex, and no one doubted their competence. In fact, as her article makes clear, they were among the best prepared and the most skillful of all the astronomers who had assembled to observe. For once, their sex could be treated as a jolly footnote rather than the defining condition of their work.

Of course, in 1869 there were very few men to compete with the Vassar astronomers. This was in part because of the lack of astronomy programs in men's colleges, but also because the Civil War had wounded and killed so many men of that generation. As the 1870s approached, the situation began to change. The new generation of men who had been too young to serve in the war was finishing college, augmented by surviving soldiers whose college careers had been interrupted by the war. Male college graduates of the class of 1870 and beyond saw professional women as usurpers of positions that were rightfully theirs. The backlash was beginning.

CHAPTER 13

The Undevout Astronomer

In her early years at Vassar, Mitchell had few allies among the faculty. She had always made friends easily, but the deck was stacked against her. It was sometimes difficult to connect with her male colleagues. Her most natural ally would have been Hannah Lyman, the "Lady Principal" and chief administrator of the school. Eventually the two did in fact become good friends, but at first there was a surprising difficulty: religion. As Maria explained it: "Miss L[yman] is a bigot, but a very sincere one. She is the most conservative person I ever met. I think her a very good woman, a woman of great energy . . . but if we had lived in the colonial days of Massachusetts, and had she been a power, she would have burned me at the stake."[1]

These were strong words, but they accurately depict the attitude of many at Vassar toward Mitchell's religious beliefs. If Mitchell had remained a Quaker, she might have encountered some difficulties at Vassar; Quakers, after all, had been tortured and even killed in colonial Massachusetts. But Mitchell was a Unitarian, and in the middle of the nineteenth century, this was considered far more dangerous than Quakerism. In fact, Mitchell's religion was radical enough that she kept it quiet in her first years at Vassar. Once, when a student's mother clung to her in tears, expressing her worry that her daughter was "not a Christian," Mitchell reported that "I was strongly tempted to avow my Unitarianism." But "Miss W., who was standing by, said, 'Miss Lyman will be an excellent spiritual advisor.'"[2]

Mitchell was wise to hold her tongue. In 1879, when she had

been at the college for fourteen years, Nathan Bishop, one of the trustees, attempted to have her dismissed because of her religion. Complaining that Mitchell was a "Rank Theodore Parker Unitarian," he wrote, "I believe she has kept away from Vassar five times as many students as her influence has drawn to it."[3] By that time Mitchell had won over Hannah Lyman, or perhaps it would be more accurate to say that the two had met each other halfway: in 1870 Mitchell wrote to Lizzie Williams, one of her beloved students from the class of 1969 who was furious at her family's conservativism, saying, "You are so over all a radical that it won't hurt you to be toned down a little. And in a few years (as the world moves) your family will have moved one way and you the other, a little, and you will find yourself on the same plane. It is much the way between Miss Lyman and myself."[4] By 1879, the other members of the board had also come around to Mitchell's tolerant way of thinking. They disagreed with Bishop that she should be fired, but none of them expressed surprised at his vehemence.

Outside New England, Mitchell's religious beliefs seemed radical, but Unitarianism was a natural choice for her. She had been raised as a Quaker, but her parents were very liberal in their attitudes, and ultimately not one of her siblings remained a member of the Society of Friends. Primed by their father's teachings, all of them explored the reformist religions that had spread across New England when they were young adults.

When Maria was a girl, William Mitchell stored all of his school paraphernalia in the attic of the little house on Vestal Street. His charts and diagrams and illustrations were tucked into the rafters above the wooden spheres he used in astronomy demonstrations, along with all sorts of detritus from his years of teaching. One artifact that the Mitchell children particularly liked was a long piece of wood painted with the slogan "An undevout astronomer is mad." When they played in the attic, they used it as a mast for a makeshift ship, which they called the *Undevout*. Throughout her life, Maria Mitchell laughingly referred to herself as "the undevout astronomer."

How did Maria Mitchell's merry undevoutness spring from the harsh gray atmosphere of the Quaker Discipline of her childhood? What did she mean when she called herself "undevout"? She certainly didn't mean that she was not devoted to astronomy, or that she was an atheist or iconoclast. But by the standard of her childhood faith, she was indeed undevout. She found spiritual matters mysterious, often unresolved, and yet somehow, along with her brothers and sisters, she found joy in spiritual mysteries rather than in the disciplined devotions of her grandparents' church. She also found a wellspring of energy that she directed toward social activism. For Mitchell, being "undevout" meant being heterodox, always ready to challenge the social status quo—and to try and change the world for the better.

Unitarianism swept like wildfire through the United States in the mid-nineteenth century. There were many reasons for the great surge of liberalization, and two were particularly important for Mitchell. First, the great changes in science required a shift in the relationship between science and faith. Second, the powerful social-reform movements of the period changed the relationship between religion and the social order. In 1800, the church was a pillar of society; most churches in America were fundamentally conservative, upholding the social order of their communities. By 1900, many churches had taken on the opposite role: sermons became social critiques as the churches plunged into political battles about slavery, then wages, and finally social and political rights. A committed Christian in early-nineteenth-century America was most likely obedient to community mores and expectations, including social hierarchies. By the late nineteenth century it was quite possible to be religious and "undevout": to believe that the natural and spiritual worlds were full of mysteries and wonders rather than simple answers and to be ready and willing to question socially accepted truths.

Louisa May Alcott described the antebellum churchgoing experience as radicalizing, writing that after church in Boston in the 1850s, "the crowds poured out; not yawning, thinking of best clothes or longing for dinner, but waked up, full of talk, and eager to do something to redeem the country and the world."[5] In the 1860s, Emily Dickinson's poems explored the new role of the "Scientist of faith," urging both scientists and people of faith to avoid dogmatism. In the 1880s Maria Mitchell declaimed, "[L]et us have truth, even if the truth be the awful denial of the good God," while expressing her hope that "scientific investigation pushed on and on will reveal new ways in which God works, and bring us to deeper revelations of the wholly unknown."[6] For Mitchell, as for many of her friends and peers, social reform and scientific inquiry shaped religious understanding. Goodness seemed a matter for reformers, while truth was the province of science. Liberal religion sought goodness and truth hand in hand with political and social action and scientific investigation, but this was an uneasy fellowship. Mitchell's hopes for revelation were never much more than hopes, and her fear that science would lead to "the awful denial of the good God" was very real. Nonetheless, she braved it out, searching genuinely, if undevoutly, for an astronomer's heaven rather than a dogmatist's.

Much earlier, in the 1820s and 1830s, the seeds of religious liberalization had been sown around Massachusetts in classrooms rather than in churches. Bronson Alcott, Elizabeth Palmer Peabody, and William Mitchell were in the vanguard of the new reformist pedagogy. In 1828, the *Nantucket Inquirer* described William Mitchell's astronomical lectures as "remarkably happy"—a phrase that holds a key to the changes in science, pedagogy, and faith taking place at the time. William Mitchell focused on the weird curves and most peculiar details of the night sky, and he did so with a glee and a sense of liberation that were utterly new to his audience.

The Mitchells wanted their children to be able to play, in part because they believed play was central to learning, but also because they trusted their children to learn for themselves. In the strict Quaker Discipline that had developed over the eighteenth century in Nantucket, such attitudes were radical. William and Lydia Mitchell were faithful and observant Quakers, but they were also ahead of their time in the liberalization of Christianity. Ultimately, not one of their children would follow in their footsteps as observant disciplined Quakers; all would strike out toward Unitarianism.

One day when Maria was young, the children lost control of the wooden planets they were playing with in the attic. It was Sunday afternoon, and they were supposed to be reading the Bible and thinking quietly about spiritual matters. Instead, as Anne Mitchell described it:

> Suddenly, a draught...set Mars...down the garret stairs, striking Jupiter as it went and Saturn flew after Jupiter. The Asteroids, little pictures with great ears, struck by the scene, rushed after the others to see

what the hue and cry meant and the Earth brought up the rear. The garret door was forced widely open, and in some strange way Jupiter received a blow which threw him with tremendous velocity down the next flight of stairs. Bump, bump! Whack, whack! Came Saturn tumbling after, whose rings seemed like so many shrieking fiddles, while popping and hopping like tiny marbles flew the little Vesta, Pallas, Juno, and Ceres, distractedly aiming everywhere.[7]

According to Maria Mitchell biographer Helen Wright, William Mitchell reacted to the hubbub with quiet cheer: "His eyes were twinkling, but his voice was stern, 'This is First Day. Have you forgotten? You will surely rouse the neighbors and perturb your grandparents next door, and even the pigs and chickens in the jail down yonder. You better rescue the sun before he burn a hole in the carpet. Then sit you down and ruminate awhile."[8]

The mind-set that brought together science, religious faith, and pedagogy within a rubric of antiauthoritarianism and self-examination is known today as transcendentalism, but it is also closely linked to Unitarianism—and both resist easy definition. These movements centered on questions rather than answers, conversations rather than sermons, and private inner truths rather than shared dogmas. Because there were no easy answers, the religious and philosophical movements in New England between the 1830s and 1870s were personally difficult for many people. William and Lydia Mitchell's children traded the certainties of their parents' faith for new uncertainties, and at times they experienced this exchange as a loss. But transcendentalism and Unitarianism also generated great enthusiasm and energy. Bronson Alcott's daughter Louisa May Alcott suffered the hard knocks of a transcendental childhood, particularly because her father's radical idealism meant constant poverty for his family. Bronson's *Conversations with Children on the Gospels,* for example, so outraged the people of Boston that his school was forced to shut down. Later the Alcotts sank their limited family resources into Fruitlands, an idealistic communal farm where they subsisted on the verge of starvation. But in spite of these privations, Louisa May Alcott celebrated her father's reformist spirit in many of her novels. *Little Women* and its sequels presented a family of girls as courageous spiritual pilgrims, determined to find spiritual truths for themselves. Her reform novel *Work,* intended for a more adult audience, included a profile of a minister she called Mr. Power, whom she based directly on the Unitarian minister Theodore Parker. When Mr. Power spoke, Alcott wrote, "religion seemed a visible and vital thing, a power that she could grasp and feel,"[9] and his teaching led directly to efforts for social reform. In *Work,* the logical consequence of liberal Christianity was feminism. At the end of her novel, Alcott described "a loving league of sisters, old and young, black and white, rich and poor, each ready to do her part to hasten the coming of the happy end."[10]

The Mitchells offer a fascinating parallel to the Alcotts. Like Bronson Alcott, William Mitchell was an accomplished educator whose teaching style was happy, peculiar, and experiential, open to the questions and insights of his students. Both Bronson Alcott and William Mitchell were somewhat well known in Massachusetts, but both were eventually outshone by the gifted second daughters whose educations they had tended so carefully. Both Louisa May Alcott and Maria Mitchell, two of the most successful women of the nineteenth century, were Unitarians who became passionately committed to women's educational and political rights. Both believed that women's right to work was paramount, that women could and would work for the good of society, and that they would change the world for the better. Their feminism was rooted in their Unitarianism.

But the Unitarianism was a mixed blessing. In 1843, when, at the age of twenty-five, Maria Mitchell informed the Nantucket monthly meeting of Women Friends that "her mind was not settled on religious subjects," she was avowing an uncertainty that would never lose its grip upon her soul. Mitchell did not leave much record of her feelings about leaving the Quakers, though she would write later in life about her religious doubts and hopes. What remain are the buildings: a Quaker meetinghouse much like the one she attended as a youngster, and the Unitarian church that she joined in its stead.

The tumultuous years from 1776 to 1865 were hard on Nantucket Quakers. Their religion prescribed pacifism, but the American Revolution, the War of 1812, and finally the Civil War were irresistible causes for many. Nantucketers who fought were excommunicated, or "read out," of the meetings. In the 1830s, the Quakers themselves divided into factions—Hicksites, Gurneyites, and Wilburites—and the once unified Nantucket Quaker community was embroiled in bitter strife. Many left altogether, choosing the open embrace of the Unitarian Second Congregational Church. After the Civil War, the island's population shrank dramatically, and the Friends' meetings withered and finally disappeared. For a long time there were no practicing Quakers on the island at all, and most of the meetinghouses were torn down. Only one Quaker meetinghouse now remains, and the Nantucket Historical Association maintains the building as part of their research library complex on Fair Street. Since 1939, Friends have once more been using the meetinghouse for weekly meetings.

When you enter the meetinghouse on Fair Street, you notice its narrowness. It is a small space, and everything about it is narrow, from the long room to the skinny wooden benches. The plainness is self-conscious. At first glance the building looks like it might be a relic of the 1680s, when the island was first settled. A casual observer would assume that the builders of the meetinghouse had few tools and little wood or money, but in fact the building was constructed in 1838, in the heyday of Nantucket whaling. The Nantucketers who built it were not poor, and they were not at all provincial. Many were cosmopolitan merchant mariners who had traveled the globe and brought back a fair share of the world's riches. The stark, even grim architecture of the meetinghouse was a self-conscious disavowal of the world and worldly things. The building imposes a strict Quaker Discipline on itself, rejecting all frivolity and coming close to rejecting all beauty.

To my eyes, there is something appealing about the stark simplicity of the meetinghouse, but mine are the eyes of a somewhat sentimental historian. I have never been subject to the discipline of an early-nineteenth-century Friends' meeting. Not only colors were forbidden; music was also frowned upon, as was too much interaction with "the world." The Discipline seemed to forbid and condemn almost everything.

On its face, the strictness of the Discipline might have turned Mitchell away, but the church was also tainted by whiffs of hypocrisy. The Quakers on Nantucket were a paradoxical lot. Many were very wealthy, and wealth and simplicity are difficult bedfellows in any circumstances. A simple Quaker dress isn't quite so simple when it is fashioned from heavy silk. The luxurious appointments that had become common on the island were at odds with the culture's purported simplicity. Moreover, the primary source of Quaker wealth was the whaling industry, a patently violent business.

Whaling was a high-stakes gamble. A few ships were lost each year, and the odds were even worse for individual men. Many whale men died ugly deaths. Those who survived their voyages came back with memories of the coastal cities of Europe and Asia, the shores of Africa, and the paradisiacal Pacific islands. Many sailors had intimate memories of island women, and even those who avoided sex with the natives had grown used to their casual attitude toward nudity. In short, whaling was un-Quaker in every way. It was a gamble. It was violent, which went against Quaker commitment to pacifism, and it was worldly in two senses: earthly, rather than spiritual, and strongly cosmopolitan. It was sexually open-minded, to say the least. Finally, whaling resulted in great wealth for the survivors, which led to luxuries that were forbidden by Quaker simplicity.

As these internal contradictions mounted, the members of the Quaker meetings on Nantucket responded by trying to tighten their grip on those who remained. In practice, this meant that many men escaped the Discipline for years at a time, while the rules got stricter for women and children. Defining itself ever more narrowly, the community "wrote people out of the meeting" left and right. When Maria Mitchell's brother Andrew was written out for marrying a non-Friend, Mitchell decided it was time for her to withdraw; the church had simply become too narrow-minded. Mitchell could not agree to all of their strictures, and she was no longer comfortable with the paradoxes and denials at the heart of the Nantucket Discipline.

Mitchell was not alone in her sentiments. Across the United States, the Second Great Awakening was flourishing; new churches and new denominations that were more open-minded, less puritanical, and less hierarchical were springing up all over. Methodists and Baptists were replacing Anglicans and Presbyterians, and the Quakers themselves were changing radically—on the mainland, the radical Hicksites were changing the orientation of Quakerism to something far more politically engaged and socially radical. Perhaps if the reformist Hicksites had been dominant on Nantucket, Mitchell would have made peace with her childhood religion, but that is unlikely. The great spiritual excitement in New England centered on Unitarianism and its philosophical cousin transcendentalism. William Channing, Ralph Waldo Emerson, and Theodore Parker led these movements, and Mitchell, like many others, found them irresistible. Their emphasis on self-reliance harkened back to the early Quaker tenets of inner light, but the Unitarians seemed to mean it: by 1868 the Second Congregational Church on Nantucket, to which Mitchell belonged, added a declaration of openness to their bylaws: "We claim no right to exclude anyone from this communion, on account of difference in doctrinal opinions."[11]

The Second Congregational Church of Nantucket stands high atop Nantucket town, overlooking the Straight Wharf. It is just a few steps from the Pacific Bank, where the Mitchells lived in the 1840s, and close to the Fair Street meetinghouse. Architecturally, it presents a great contrast to the Quaker building. It is four or five times larger, a grand, open space with light pouring in through high windows. The ceiling soars above the room. It not elaborately decorated; the walls and ceiling are painted with trompe l'oeil pillars, but the plasterwork is flat and plain. The defining qualities of the space are size, light, height, and generosity—it is anything but narrow or confined.

Even so, the confinement of any church was nearly too much for Maria Mitchell. She rented a pew and paid her dues faithfully, even long after she had moved from the island, but the pew she chose, number twenty-two, is the back pew in the far right corner—the most private, least committed location in the sanctuary. Perhaps on some Sundays in summer all of the pews were full; otherwise, I imagine a cluster of worshippers close to the front, then a series of empty rows, and then Mitchell alone at the back, observing but refusing to be drawn in too far. Her seat would have taken great advantage of the acoustics of the room—it was a wonderful place to listen and observe, but not necessarily to participate.

The covenant of the Second Congregational Church was adopted in 1837. The congregants agreed to model their covenant on the Unitarian covenant of Harvard University—in the words of Henry Edes, pastor of Nantucket's Second Congregational Church at the time, "We believe that the corruption and sin into which men are plunged are their own work, that reason and conscience and power of choice, and will, were given them, and are sufficient to enable them both to understand and to obey their duty."[12] Refusing to believe in Original Sin, they focused instead on reason, conscience, choice, and will, creating a religion of individual souls and the human intellect. Nantucket Unitarians believed in the power of the human mind. Some thought that God had given humans their intellects; others were less interested in God per se. All believed that human intelligence could work for goodness: could determine what was good, choose it over evil, and work toward it. Perhaps put off by the bitter sectarian battles that had rocked the Quakers, the Unitarians were chary of making specific pronouncements about personal morality, and even hesitant to make specific declarations about theology. Political and social reform, yes; policing of individual conduct, not so much. The Unitarian Church was the perfect spiritual home for someone as conscientious and as intelligent at Maria Mitchell.

In the decades leading up to the Civil War, Massachusetts was rocked with social and spiritual upheavals. Abolition, women's rights, socialism, and modern science were all stirred together into a heady spiritual stew that Theodore Parker framed as Unitarianism. All over the state, liberals left Congregationalist churches and Quaker meetings behind. Many of them left because the great paradigm shifts of the new science shook the foundations of their faith, often causing genuine anxiety and despair. The ideas of Copernicus, Galileo, and Newton decentered the cosmos, and for some nineteenth-century thinkers, these ideas dethroned God. Ralph Waldo Emerson explained this publicly by writing, "Astronomy taught us our insignificance in Nature." Privately, he mused, "I regard it as the irresistible effect of the Copernican astronomy to have made the theological scheme of redemption absolutely impossible."[13] As an astronomer, Mitchell wrestled with the same questions. After a long and painful spiritual search, she left the certainties of the religion that had fostered her. Instead of a Quaker, she decided to be a scientist and a reformer.

Many years later, Mitchell's wide circle of Unitarian friends would provide many of Vassar's most distinguished guests. The students were thrilled when Louisa May Alcott visited the college. Frances Wood, the college librarian at the time, recalled:

> Every woman speaker of note in that day, Julia Ward Howe, Anna Dickinson, Mary Livermore, Elizabeth Cady Stanton, Ednah Cheney, were personal friends of Miss Mitchell, and at various times her guests at the observatory. Sometimes one would be invited to speak before the whole college, as when Mrs. Howe recited her Battle Hymn of the Republic and Mrs. Livermore told the story of her own experiences in a hospital. But the observatory was chiefly the place of meeting, with spirited talk and free discussion. What delightful evenings were there when a favored few were asked to meet distinguished guests! What personal anecdote and reminiscence! And what good coffee at the end!...

In the years after Mitchell's father died, she gradually changed her Sunday suppers from the student-centered celebrations that he had loved so dearly to a more thoughtful gathering of Unitarian-leaning members of the Vassar staff. Frances Wood loved these evenings:

> There were only four or five of us outside the students who had Unitarian preferences, and we were considered, on account of this, by our more orthodox sisters to be indeed very black sheep religiously. The club was very informal, meeting every Sunday evening for supper at the observatory, or in the room of a member, entertaining in this way any guest visiting at the college that Miss Mitchell considered in sympathy enough to invite. The aim was serious, excluding all gossip and light talk and while there was no straining after "high thinking," it was certain we had a frequent hint of this in Miss Mitchell's independent, stimulating opinions, and from the friends who came as guests, bringing their best thoughts with them.... She had a deeply religious nature.[14]

Perhaps because of her deeply religious nature, Maria Mitchell didn't stop searching. From 1846, when she left the Quakers, she never rested firm in any particular belief, but she never stopped thinking about heaven. In her later years at Vassar, Mitchell put some of the agonies of her earlier spiritual doubts behind her, as she had come to believe in doubt itself. She told her students, "We cannot accept anything as granted beyond the first mathematical formulae. Question everything else."[15]

Mitchell loved the simple clarity of equations. Her student and successor Mary Whitney remembered Mitchell saying, "A mathematical formula is a hymn of the universe." Mitchell wrote in her notebook, "Every formula which expresses a law of nature is a hymn of praise to God." Whether they praised God or the universe, formulae for Mitchell were aesthetically beautiful, like poems. And they weren't just any poems—Mitchell experienced mathematical formulae as songs of praise and celebration. For the nonscientist, it can be hard to understand the aesthetic beauty or the eternal significance of an equation. Part of the charm lies in the power of an equation's simplicity, or the elegance of its symmetry. Newton's simple inverse-square law, for example, expresses the complicated gravitational relationship between the earth and the moon in just a few symbols. It also explains the changing intensity of your cell phone signal as you drive away from a broadcasting tower and the dimming at the edges of the pool of light around a streetlamp. That one simple mathematical formulation—the inverse square—can capture the shifting intensity of gravity, electromagnetic fields, and even light is nothing short of miraculous. But gravity, light, and even cell phone signals are rather quotidian miracles. None are overtly spiritual; none give explicit clues of what God is like. The more we learn about the universe, the less we seem to know about the heavens, a fact that sometimes frustrated Mitchell. She wrote that "the world of learning is so broad and the human soul so limited in power! We reach forth and strain every nerve, but we seize only a bit of the curtain that holds the infinite from us."[16]

By the 1880s, Mitchell had become a stout defender of the uncertainties of scientific inquiry. She refused to think of scientists as enemies of religious faith or religious truth; instead, she described scientists as seekers of truth. In July 1883, she asked,

... can the study of *truth* do harm? Does not every scientist seek only to know the truth? And in our deep ignorance of what is truth, shall we dread the searching after it?

I hold the simple student of nature in holy reverence and while there live sensualists and despots and men who are wholly self-seeking, I cannot bear to have these sincere workers held up in the least degree to reproach. And let us have truth, even if the truth be the awful denial of the good God. We must face the light and not bury our heads in the Earth. I am hopeful that scientific investigation pushed on and on will reveal new ways in which God works and bring us to deeper revelations of the wholly unknown. The physical and the Spiritual seem at present separated by an impassable gulf, but at any second that gulf may be over-leapt, possibly a new revelation may come.[17]

For Mitchell, the mission to "face the light" and "seek only for the truth" was urgent. She hoped for great revelations in time—a chance to leap over the gulf that seemed to separate the physical from the spiritual.

CHAPTER 14

Retrograde Motion

After the Civil War, the United States was desperate for educated workers. The war had literally decimated a generation of men. Vassar, Wellesley, and Smith colleges all opened during the brief window when women's higher education seemed to be in the national interest.

Then in 1873, as the first unscathed generation of young men finished college, women's colleges came under fire. Edward Clarke, a professor at Harvard Medical School, argued in a short book called *Sex in Education* that college-level education was dangerous to women's physiology. Describing the uterus as a delicate organ with a voracious thirst for blood, Clarke argued that college women thought too much and their highly charged brains drew too much blood out of their systems, eventually causing their sexual organs to shrivel away. Education was, he explained, a "sterilizing influence" for women.[1] Strenuous thinking could cause women to substitute "masculineness for distinctive feminine traits" and to risk becoming sterile, unwomanly creatures, "analogous to the sexless class of termites."[2]

In Clarke's circular antifeminist argument, women who argued for political or educational rights were those who had already "drifted into an hermaphroditic condition,"[3] while only the woman who had "become thoroughly masculine in nature, or hermaphroditic in mind"—who had "divested herself of her sex"—could take a man's "ground and do his work."[4]

Although Clarke argued against all higher education for women, the scientific basis of his argument had the strongest implications for women in the sciences. Cynthia Russett, the Yale historian, includes Clarke's work among many examples of Victorian "scientific antifeminism."[5] She comments that this attitude "needs to be seen for the masculine power play that it was, but it needs to be seen also as an intellectual monument, etched in fear, of the painful transition to the modern world view."[6] Late-nineteenth-century (male) scientists tried frantically to establish a bright line that would offer a clear division between the sexes, to find a physiological basis for "true sex." This was a surprisingly difficult task. Human anatomy is varied, and some people are anatomically hard to classify. Before chromosomal analysis was available, scientists studied anatomical structures and differences in gonadal tissues, searching desperately for a scientific way to define the differences between men and women. When physiology proved inadequate, scientists turned to psychology, but this was even murkier.

The historian Michel Foucault explains late-nineteenth-century attitudes as part of a quest to define or even establish "true sex."[7] One side effect of this quest was that "true" women were defined by scientists as intellectually inferior and barred from scientific institutions. There were many reasons for excluding women from the sciences, but among the most significant were male scientists' fragile sense of their own professionalism and their anxiety about broader social upheavals. Scientists also had a paradoxical impetus to exclude women: the implications of many scientific theories threatened to diminish masculine certainty in painful and frightening ways. What this meant in practice was that science as a discipline was redefining itself as hostile ground for women.

Maria Mitchell strongly objected to Edward Clarke's arguments. When Edward Everett Hale published a laudatory review of Clarke's book in his magazine, the *Christian Examiner* (known as *Scribner's* after 1875), she wrote to him personally in protest. Her letter does not survive, but Hale's response does. He wrote,

> The questions which relate to the education and work of women have been so long conducted . . . as if they were only matters of sentiment, that Dr. Clarke's book seems to have excited unmercifully surprise. I am

sure that you do not share that surprise. I am sure that you believe that a mention of sex must be discussed as a question of sex. I have it in my power to say that questions of sex shall not be discussed in my magazine. Thirty years ago, I probably should have said so. It is not I or people who think as I do that have brought these prominently forward into general conversation and argument. Now that they are so brought forward I think it would be absurd for old and new to decline to recognize them.[8]

Hale's defense is rather lame, but his letter illustrates the widespread discussion of Clarke's arguments. Defenders of women's education were put in a very awkward position. Menstruation, for instance, was central to the argument against women's higher education, so the only way to counter Clarke was to try to gather some facts about college girls' menstrual cycles and general health and keep follow-up records of college graduates' subsequent fertility. Education had become "a question of sex."

In December 1873, shortly after *Sex in Education* was published, Maria Mitchell ventured to Boston. She was greeted with questions about sex, and particularly the sexual health of her students. She wrote to John Raymond, the president of the college:

> My dear President,
>   No sooner am I in Boston than I have to begin the battle for the College!
>   Dr. Derby, prominent physician in the city, asked me about the number of students who graduate before nineteen—about the number who are hurt by rooming on the fourth floor—&c &c with medical details of his patients from Vassar, not particularly pleasant to hear.
>   I combated the implied fault-finding at the physical training at Vassar until I was pretty well tired, and coming out of the office met at once a reporter, Miss Joy, who asked me if she would come up to Vassar with a view to writing a letter for the Boston Post. I asked her to come as my guest, when the holidays were over. . . . It does seem to me desirable that whether we are right or wrong at Vassar, those who will write about us should know what Vassar is.[9]

Mitchell was not alone in feeling embattled, but as the leading example of educated American womanhood, she was under particularly harsh scrutiny. Each inquiry about her students' uterine functions was also an indirect inquiry about her own. If education unsexed women, turning them into hermaphroditic termites, then Mitchell, a fifty-five-year-old postmenopausal female college professor who had recently received an honorary doctorate (from Rutgers in 1870), was no longer a woman at all. Instead, she was an unsexed figure whose life's work was to drain young girls of their womanhood. The charges were upsetting, to say the least.

Edward Everett Hale invited Mitchell to respond to Clarke in the pages of his magazine, but instead she encouraged her colleague Alida Avery, the Vassar College physician and professor of physiology, to write an essay for Julia Ward Howe, who was compiling a book of essays that refuted Clarke. Howe's anthology *Sex and Education: A Reply to E. H. Clarke's Sex in Education* was published the following year, in 1874. Avery's careful exposition of the medical records of Vassar students from 1865 to 1874 was irrefutable evidence that Clarke's theory was baseless. Mitchell's own responses would come in essays she published and lectures she delivered in her classroom and around the country over the next few decades.

Allies in the fight against Clarke, Julia Ward Howe and Maria Mitchell soon became close friends. In New York in 1872 they both attended the first Women's Congress, a national conference that brought together those working for women's educational, political, and labor rights, and when Mitchell agreed to serve as president of the resulting Association for the Advancement of Women, Howe enthusiastically signed on as her vice president. Howe was a prolific writer—most famous for the "Battle Hymn of the Republic," she also published a number of philosophical essays, a travel book, and four volumes of poetry, and she'd written unpublished plays and a secret novel. In the introduction to *Sex and Education: A Reply,* Howe declared, "The philosophy of sex is thus far little understood in America, or anywhere else."[10] Howe spent more than fifty years trying to work out a philosophy of sex in her own writing, most notably in her novel about a hermaphrodite name Laurence, which was not published until 2005. In 1883 she published a biography of Margaret Fuller, and in 1884 she wrote the first biography of Maria Mitchell, a twenty-five-page chapter in *Our Famous Women.*

Although the two women had met each other a few times before the 1870s, it was their shared opposition to Clarke that brought them together. Both women were touched to the core by Clarke's claim that educated women were sexually deviant. Mitchell may have been hurt by the implication that she was sexually perverse, since her closest bonds had always been with women, but she also probably shared Howe's more philosophical questions about whether or not an intellectual woman was some sort of spiritual hermaphrodite. Although both women struggled privately with these "questions of sex," and Mitchell even went so far as to comment in her diary, "Sometimes I am distressed for fear Dr. Clarke is not so far wrong,"[11] both stoutly defended the cause of intellectual women.

Howe was also a loyal defender of Mitchell, who was personally implicated in much of the innuendo of the unsexing debate. She described Mitchell as "a gifted, noble woman, devoted to science but heart-loyal to every social and personal duty." Celebrating Mitchell's leadership of the Association for the Advancement of Women, Howe wrote, "Her customary manner had in it a little of the Quaker shyness, but when she appeared on the platform the power of command, or rather of control, appeared in all she said or did. In figure she was erect and above middle height. Her dress was a rich, plain black silk, made after a plain but becoming fashion. The contrast between her silver curls and black eyes was striking."[12] Howe's descriptions of Mitchell reveal her great affection for the friend whom she called a "sister planet,"[13] but they must also be understood as the first of many defenses of Mitchell. Howe understood astronomy well, so when she claimed "sister planet" status with Mitchell, she knew that her metaphor inscribed a path with many reverses.

As the backlash against the higher education of women gained force, Mitchell's public speech and writing grew stronger in response. In 1874 she made a speech before the Congress of Women on the "Higher Education of Women," in which she lamented what she saw as the educational apathy of the 1870s: "I rarely meet a woman in my own country who is interested in the higher education of women, unless she herself is an educator, and the mass of our people do not believe in the education of women."[14] She believed that "when the American girl carries her energy into the great questions of humanity, into the practical problems of life; when she takes home to her heart the interests of education, of government and of religion, what may we not hope for our country!"[15]

In her 1875 presidential address to the AAW, Mitchell made the case for educated women, and particularly for women in the sciences. "In my younger days, when I was pained by the half educated, loose, and inaccurate ways which we all had, I used to say, 'How much women need science,' but since I have known some of the workers in science who were not always true to the teachings of nature, who have loved self more than science, I have said, 'How much science needs women.'"[16] Mitchell's resounding declaration that science needed women would be a rallying cry for her followers throughout the last decades of the nineteenth century. Mitchell thought women were more perceptive than men and were better at the collaborative work necessary to scientific endeavors. She also saw them as less selfish. She had come to believe that Dr. Clarke willfully ignored the evidence and overlooked the "teachings of nature" because of masculine self-interest. Women, on the other hand, could be trusted to be fair and to build their arguments on concrete evidence.

Mitchell thought of the Association for the Advancement of Women as a "social science" organization and believed that the most effective strategy for advancing women's causes was gathering hard data about their circumstances. "[A] solid phalanx of figures," she told the group, "is a formidable opponent to a flourish of rhetoric."[17] For the next decade she would chair the AAW's committee on science and would do her best to gather specific information about the lives and working conditions of scientific women in the United States.

The next step was rallying the troops and instilling a generation of young women scientists with confidence and purpose. Along with the phalanx of figures, she also came up with a few rhetorical flourishes. Perhaps her most inspiring lecture to her students was the brief one called "I Am But a Woman," written in 1874:

> "I am but a woman!"
>
> For women there are undoubtedly great difficulties in the path, but so much the more to overcome. First, no woman should say, "I am but a woman!" But a woman! What more can you ask to be?
>
> Born a woman—born with the average brain of humanity—born

with more than the average heart—if you are mortal, what higher destiny could you have? No matter where you are nor what you are, you are a power—your influence in incalculable; personal influence is always underrated by the person. We are all centres of spheres—we see the portions of the sphere above us, and we see how little we affect it. We forget the part of the sphere around and before us—it extends just as far every way.[18]

When Mitchell talked to her students this way, she used the logic of spherical astronomy and the language of gravitational power to describe the personal power she believed each person could claim. Mitchell wanted her students to believe, with her, "There will come with the greater love of science greater love to one another." The science she passionately believed in was generous in spirit, optimistic in outlook, and loving of heart.

But these decades were the heyday of "scientific antifeminism," a collection of theories and practices that began with Clarke's unsexing theories and snowballed through the last decades of the nineteenth century, ceaselessly belittling women and their intellectual and physical capacities. Mitchell was fighting a losing battle. By 1885, her last year as chair of the committee on science for the AAW, Mitchell would mournfully write in her annual report, "In the half-lighted and wholly unventilated offices, women work patiently at the formulae, and pile up the logarithmic figures; in the open air, under the blue sky or the starlight canopy, boys and men make the measurements.... There seems to have been a backward movement."[19] Mitchell's sad awareness that women had moved backward in the sciences, from original researchers to assistants, was a bitter conclusion to her long career. After the 1885 report, she published no more.

Every morning in Mitchell's first decade at Vassar, she covered her cot in the observatory clock room with a velvet spread and pillows, transforming it into a lounge in order to efface the traces of her sleeping body from her parlor, study, and classroom. In the evening, and sometimes long into the night, she hosted students who wanted to observe the night sky or talk about astronomy or whatever else was on their minds. When the students went home, Mitchell went to work on her research, either making observations or calculating. Very late, she fell asleep on the narrow cot.

Then, in 1875, the closet that had served as the observatory's coal bin was fitted up as a bedroom for her. The Vassar student paper published an article, and a student wrote a commemorative poem:

> Beautiful Venus, pride of the morning
> Tell it to all little stars who have fled
> That in a sweet chamber that needs no adorning
> Miss Mitchell sleeps in a bed![20]

Miss Mitchell was now fifty-seven. She held honorary doctorates from Rutgers University and Smith College. She was a member of the American Philosophical Society and the American Academy of Arts and Sciences. She had been the computer of Venus for the United States *Nautical Almanac* for nineteen years, and she had been the first American astronomer awarded the gold medal from the king of Denmark that honored her discovery of a comet. She had served as president of the Association for the Advancement of Women. And after a decade on a couch in a classroom, she'd been promoted to a bedstead in a former coal bin.

These domestic circumstances are outrageous, but it's important to remember that they were absolutely appropriate at the time. A professional woman scientist was not a domestic being. She had no need for, or right to, the domestic sphere. When Mitchell chose to accept the professorship, she had also chosen to deny the reality of her body, to renounce domestic comfort. In short, she was no longer a woman, as the nineteenth century defined the term. Mitchell was officially unsexed.

How did Mitchell feel about her move into the closet? Perhaps it was a small victory after ten years of asking for accommodations more in line with the houses provided for each male professor, which came complete with firewood and servants. But perhaps Mitchell felt outraged that she was being forced into a closet. Perhaps Vassar College was afraid that Mitchell's visible sleeping quarters would fuel public suspicions about women's education and sexual deviance. Perhaps, moreover, it hurt her to get the message that her aging female body must be concealed.

Most of us like to think that history is fundamentally progressive, that things improve and advance as time passes. Sometimes the direction of historical movement is not so clear. Mitchell's life story can be interpreted in various, sometimes contradictory ways. In some ways, hers is the story of a progressive heroine who pushed forward for herself and all women; but when she started out, in the 1830s, American attitudes toward education for women were positive and getting more so. Women were generally encouraged to enter a variety of disciplines. By the time Mitchell was in a position to encourage others, women's education was under attack, and a cultural consensus was forming that science was particularly inappropriate for women. For the next hundred years, from 1875 to 1975, women were shepherded into the "soft" fields of the humanities.

As the years passed, Mitchell's little sleeping compartment may have come to seem like a perfect symbol for the backward movement that beset scientific women in the late nineteenth century. Women were losing access to the "starlight canopy" and being forced into "half-lighted and wholly unventilated" rooms. The great possibilities of the early nineteenth century seemed further and further out of reach—not just for Mitchell but also, and perhaps more important, for the generations that would follow her.

The last decades of the nineteenth century were decades of backward social movement for many people, not just women in the sciences. Sometime in the 1860s, Mitchell had concluded that women's political, educational, and employment rights could not be separated from one another, that the spirit of the age demanded that educational opportunity be tied to political rights and to employment reform. Only the law could protect women from being exploited by their employers, so public policy needed to enforce fair working conditions and fair wages for women, and therefore women needed to be involved in making public policy. At the same time, education was a necessary stepping-stone to women's self-sufficiency and employment rights. Only educated, qualified women could lay claim to professional jobs and to higher-income jobs; without education, women were doomed to second-class economic status. For Mitchell, the three causes were part of a single movement.

For some late-nineteenth-century reformers, the movement was as much about racial equality as it was about gender equality. Frederick Douglass and Lydia Maria Child, for example, always linked the two. But, along with most white women, Maria Mitchell made a strategic decision to focus on white women's rights and to distance herself from the struggle for racial equality.

Starting with Mitchell's generation, the American feminist movement became a white, middle-class movement; between 1875 and 1975, white feminists made little effort to find common cause with women or men of color. The strategy was racist, and it was also entirely ineffective. Most white feminists thought of black men as rivals and competed against them for political rights; many black leaders, in turn, used the rhetoric of manliness and the threat of women's rights to try to solidify black political rights. Both groups lost far more than they gained. If the pre–Civil War coalition had held, white women and women and men of color might have gained rights much more quickly, and the United States could have emerged from the 1860s a more egalitarian country. The era of Jim Crow, which stretched for a hundred years starting in the late nineteenth century, was arguably even more racist than the slave era that preceded it, as lynch law held the Jim Crow states in its cold grip. Although black men got the vote in 1870 and women of all races got the vote much later, in 1920, social mobility and educational opportunities for both groups actually declined by most measures between 1875 and 1950. By the 1960s, Betty Friedan's study of Vassar students found them far behind the women who had gotten Vassar degrees a hundred years earlier. Few intended to work; most saw their educations as a social affair. The vast majority felt they were being indulged in a few years of thinking before marriage and that there was great social pressure to forgo intellectual activity in favor of domestic activity. By 1960, although the feminization of early education was in full force, higher education had come to be seen as unfeminine. It was a threat to domesticity.

Mitchell had been opposed to slavery when she was young. She was proud of her friendship with Frederick Douglass; in the last year of her life she made the effort to go to a reception held in his honor on Nantucket, and carefully saved the invitation in her scrapbook. She had no desire to be racist. But in a harshly racist and sexist time, when rights for most people other than white men seemed to be moving backward, Mitchell quietly acquiesced in the racism of the white feminist movement and of Vassar College itself. Vassar did not consciously admit a student of color until 1940, and Mitchell never objected to this policy. She was caught up in her own painful struggle, shut in a closet by night, unable to find the largeness of view or of spirit that had been hers in the years before the war. Her last years at Vassar were bitterly sad.

In 1875, Martha Goddard wrote an angry letter to a friend refuting the "Vassar libel": the growing public suspicion that Vassar women were either unsexed by their education or oversexed by it, lesbianized by their years in a female community. The libel was painful to Maria Mitchell, but it was even harder on her students. Women who wanted to study science found more and more obstacles in their paths, beginning with the opposition of their families. In October 1874, Julia Pease wrote to her father, "I presume you will be very glad to know that I have natural Philosophy instead of a second year of Astronomy... Knowing that you... are not very desirous that I should study the higher mathematics, I have made the change.... Miss Mitchell was kind enough to say that she missed me from the class, and invited me to go over and use the instruments whenever I desired."[21] Pease's regretful obedience here is striking, as is her father's anxiety about advanced mathematics. The same year, Pease's classmate Mary Gaston wrote a letter to her mother with a bold superscript: "DON'T READ THIS TO PAPPA." She went on to confide, "I have got a splendid picture of Miss Mitchell...; you mustn't show it to Pappa, because he called her an old witch, and she is just lovely."[22] Neither Pease nor Gaston pursued astronomy or the sciences very far; in light of their fathers' active, even hostile discouragement, their decisions to focus elsewhere seem almost inevitable.

There is no way to calculate how many women students regretfully decided to give in to their fathers' tastes and interests at the expense of their own curiosity about science. Some stayed the course, though they tended to wax apologetic about it. Caroline Furness, who had been a student in the last year before Mitchell retired, wrote to inform her father that she had decided to go to graduate school in astronomy: "Now dear Papa, do not think I am selfish in planning so for myself. I must make my own way in the world. I have ability and interest, and why should I not rise to as much distinction in my profession as any man. If I were your son instead of your daughter, you would fully approve my ambition."[23]

Caroline Furness would have a hard time rising to distinction in her profession for manifold reasons. First, it was almost impossible for a woman to get a doctorate in astronomy. Second, there were still few jobs in this field. Women were hired as computers at some observatories, but they were discouraged from observing or offering their own interpretations of results. Another possibility was teaching astronomy in a women's college, but these jobs were few and far between, and they had pitfalls of their own, as Mitchell's experience had shown. It was hard to focus on research while attending to the intense demands of teaching, and it was rare to find an administration that supported women faculty in scientific research. In many ways, late-nineteenth-century women's colleges hewed, even if unconsciously, to Milo P. Jewett's model of considering women faculty "teachers" rather than professors. In spite of the obstacles, Caroline Furness eventually reached the pinnacle of astronomical success when she became director of the Vassar College Observatory. Although she had taken her first astronomy class under Maria Mitchell, Furness was not so much a student of Mitchell's as a "grand-student," to use the historian Margaret Rossiter's phrase. Her primary mentor was Mary Whitney.

Mary Whitney was one of the original Hexagon of astronomy students who had made Maria Mitchell's first years at Vassar such a delight. She graduated from Vassar in 1868 and then studied at Harvard as a "special student" in 1869 to 1870. Whitney moved on to Zurich, where she studied advanced mathematics for two years, and then returned to Harvard in 1872 for another year as a special student. From 1872 to 1881, she cast about for astronomical employment, teaching for two years in a Boston high school. Then, in 1881, she returned to Vassar. At first she worked as an unpaid assistant to Mitchell; eventually she was appointed professor; and when Mitchell retired, in 1888, Mary Whitney was her successor.[24]

In a sense, Whitney was the lucky one. It is perhaps because she had been the most talented of Mitchell's students in the first year that she was eventually able to secure a position at Vassar. But there was only one such job at the college. The young women that Mitchell trained in advanced astronomy after Mary Whitney were sent out into a field that was growing less and less hospitable to them.

In 1896, another student of Mitchell's, Antonia Maury, class of 1887, wrote a poem commemorating her days at the Vassar College observatory. She remembered:

> A plain but ample workroom
> Where a clock on marble piers
> With low beat follows the stars
> That trace the slowly moving years
>
> And all that stay in that chamber
> Or cross but its stone threshold
> Whether for use or knowledge
> All things in common hold.
>
> ... The gifts of earth and heaven
> I gathered and brought them home
> And these infinite things of Nature
> Found space in that low room.
>
> For a nameless light and freedom
> Lifts the rude walls away
> And Nature blends with the spirit
> In love's dissolving day
>
> And the tower I find it thus ever
> As I knew it in days gone by
> For dome was over it never
> Except the endless sky.[25]

Antonia Maury had good reason to mourn the "endless sky": from 1888 to 1933, she worked in the "women's department," that "half-lighted and wholly unventilated room" in which women toiled at the Harvard Observatory. During that time she made some significant contributions to variable star astronomy, but her chiefs discouraged her from thinking about astronomy—they wanted her to stick to analyzing thousands of photographs, visually scanning them for minute differences. Dorrit Hoffleit relates, "She was the most original thinker of all the women Pickering employed; but instead of encouraging her attempts at interpreting observations, he was only

irritated by her independence and departure from assigned and expected routine."[26]

Antonia Maury and Mary Whitney held the two possible jobs
for women astronomers in the late nineteenth century: Maury was
a "cheap assistant" who spent her days in a small room doing the
routinized factory work of variable star astronomy, measuring her
accomplishments in kilo-girl-hours, as one of her bosses, Harlow
Shapley, jovially put it. Whitney escaped the dullness of routine
astronomical computing, but she too had a difficult time doing original research while teaching a growing number of students. Nonetheless, she persevered. She wrote that students used the telescopes
"up to ten o'clock: then Miss Furness and I begin our special work
and carry it on until 1 or 2 a. m."[27] Whitney and Furness were following in Miss Mitchell's sleepless footsteps, and eventually their
self-sacrifices began to add up. "A bibliography of the publications
of the staff of Vassar College Observatory while Whitney was director lists 102 articles and other publications." This was "the largest
amount of scientific work of any of the women's colleges."[28]

Beyond astronomy, two of Mitchell's students gained fame in
other scientific fields. Ellen Swallow Richards, class of 1870, was a
special student in chemistry at MIT in the 1870s, and eventually
became an instructor there. Richards specialized in domestic chemistry, which came to be known as home economics. Christine Ladd-
Franklin, class of 1878, was a special student at Johns Hopkins,
starting in mathematics and eventually completing all of the requirements for a PhD in experimental psychology.

There are some interesting parallels between Richards and Ladd-
Franklin. Both women were admitted to graduate programs as "special students," as Mary Whitney had been at Harvard. All were
exempted from paying tuition. Later, it became clear that the universities had been careful not to allow this first generation of women
students into their official records: granting them free tuition
allowed the administrations to deny that they had ever had any
women students. Margaret Rossiter analyzes the "special student"
strategy of the 1870s and 1880s as a consciously exclusionary policy.
None of these women received PhDs, not even honorary doctorates
like Mitchell's. The ever-combative Christine Ladd-Franklin completed the requirements for a PhD at Johns Hopkins, but the degree

was summarily denied. As far as the university was concerned, she had never officially earned it.

Another interesting parallel between Ladd-Franklin and Richards was that both were diverted from "hard" sciences toward softer ones. Ladd-Franklin taught experimental psychology as a hard-edged statistical discipline in her adjunct position at Columbia University for many years, and although she was excluded from psychologists' associations, she continued to hammer away at them, pushing for harder science in psychology and battling for herself and for other women to be admitted to the disciplinary conclaves.

Richards took another approach. She turned from chemistry to home economics, arguing that science could and should be womanly and that the science of the home was the most important for women to study. Taking a nonconfrontational approach in line with prevailing domestic ideologies, she tried to use domesticity as a point of entry for women in the sciences. In some ways, she was much more successful that Ladd-Franklin: home economics flourished in women's colleges and in coeducational colleges and universities across the country. But this was arguably a Pyrrhic victory. Learning to sterilize baby bottles and canning jars was mundane material for college coursework, and a surprisingly dull way to use a laboratory autoclave. Domestic science, as Ellen Swallow Richards taught it, was bitterly far from the infinite expanses of the starry sky and mathematical rigors that her teacher Maria Mitchell had shown her.

The last years of the nineteenth century were harsh ones for women in the sciences. In one way or another, each of Mitchell's students was hemmed in, forced into an ever smaller space, and eventually left with nothing but memories of intellectual freedom. Antonia Maury's lament expressed the longing of many former Vassar students:

> For a nameless light and freedom
> Lifts the rude walls away
> And Nature blends with the spirit
> In love's dissolving day.[29]

CHAPTER 15

Urania's Inversion

For most of her life, Maria Mitchell was celebrated as America's sole scientific heroine. She received endless accolades for her independent mind, her strength of character, and her intellectual abilities. In her final years, there was more controversy around the scientific education of women, but there were very few public attacks on Mitchell herself. Those that surfaced were quickly expunged from the record by Mitchell's loyal network of friends, colleagues, and sisters. But some scholars argue that Mitchell's celebrity was also a form of control by those who created it. Singling her out as the only exceptional woman scientist and magnifying her virtues to divine proportions served to distance her from other, more ordinary women, and helped to make her opportunities and achievements seem out of reach. Celebrity for Mitchell was a double-edged sword. She tried to wield it on behalf of all women's intellectual culture, working indefatigably though her fifties and sixties for the good of all scholarly women, and scientific women in particular.

An embodiment of Yankee grit and determination, Mitchell had a brusque wit and was always ready with a hard-edged comeback. When she heard from a student that John Raymond, the president of Vassar, thought she criticized him too much, she wrote him a quick, mocking note, assuring him that she did not tell students all of his faults because she did "not know all, but if I undertook to tell all that I do know, it would take all the time I have!"[1] Mitchell was also known for her disdain for domesticity. When a visitor suggested that she set up a little kitchenette in the observatory, she angrily asked him, "Is my time worth nothing more than to boil eggs?" and refused to talk to him any further. She preferred to grab a bite at the college dining hall, or to subsist on apples and crackers sent from home. Mitchell bragged about her simple living, writing in her invitation to Dorothea Dix that "my accommodations are more scientific than domestic, but I shall endeavor to make you comfortable."[2] She also railed her entire life against sewing. After she visited a cooking school in 1886, she observed in her journal, "Do not so fill the day with bread and butter and stitches that no time is left."[3]

What Mitchell valued was original thought. During her twenty-two-year career at Vassar, her most constant advice to her students was to work hard and think for themselves. She asked students, "Who settles the way? Is there anyone so forgetful of the sovereignty bestowed on her by God that she accepts a leader—one who shall capture her mind? There is this great danger in student life. Now, we rest all upon what Socrates said, or what Copernicus taught; how can we dispute authority which has come down to us all established, for ages? We must at least question it."[4] But in spite of her own taste and preferences, as John Lankford puts it, "Maria Mitchell as myth and symbol was interpreted to fit changing cultural needs. Post Civil War women moved, for the most part, away from reform and toward domesticity. Thus, the hard edges of the image (Maria Mitchell as competent scientist and independent woman who fought for equal pay) had to be softened."[5]

As cultural attitudes changed, Mitchell did too; she was starting to age. In 1879, when she was sixty-one, the *New York Times* gushed, "Miss Mitchell is also lauded as a delightful acquisition socially, still retaining the warmth and freshness of youth, in spite of the heavy burden of cares and honors. She is said, on her annual vacation return to Nantucket wind and waves, to throw these off as a superfluous garment, and become again a rollicking school-girl, rejoicing in her freedom, and laughing and shouting in her gambols with Neptune among the merriest of them all."[6] But the years were beginning to take their toll, especially when she was far from Nantucket. On April 30, 1882, she grumbled in her journal,

I am very well, but age tells on me. My feet are lame if I wear old shoes, my new teeth make my gums sore. It is useless to try to console myself with the recollection that when I was young, new shoes harassed me and old teeth kept me awake; the pains of youth are easily forgotten and quickly remedied; those of age cling to you and must be borne. I am thankful to have nothing worse of physical ills.

Emerson's death is a grief to me. He would have done no more for the world, but it was good to see him and rarely to hear his voice. One could have wished to have lived his life. I have rarely felt this, but I did when young, for Dr. Bowditch [the navigator], later for Mrs. Somerville, and now for Emerson.[7]

All of her role models had gone before her, but still she continued to teach and to enjoy her life at Vassar. That spring, she scrawled, "Vassar is getting pretty. I gathered lilies of the valley this morning. The young robins are out in a tree close by us, and the phoebe has built a nest, as usual, under the front steps. I am rushing dome poetry, but so far show no alarming flashes of brilliancy."[8] With classes, the occasional lecture, visits from luminaries and old friends such as John Greenleaf Whittier and Matthew Arnold, and pleasant rituals such as the dome parties, the years flew by.

In January 1888, Maria Mitchell quite suddenly retired from Vassar. She still loved her work but had found herself too weak to go on. She wrote to the college president, "My more than half-century of work has worn and tired me; my physician advises rest."[9] A few days later, she left Vassar for good. That winter, the board unanimously appointed her to a salaried emeritus professorship as a token of her extraordinary service to the college. She moved back to her own house in Lynn, where she could be near her sister Kate Mitchell Dame, and where she could construct another observatory with the telescope that had been given to her so long ago by the "women of America." Relieved of the responsibilities of the college, she rallied, and in the spring of 1888, she wrote to her friend Mary Raymond (the wife of John, the former Vassar president) to report that she was embarking on a new study. "I am studying Greek!" she proclaimed. "It will take 30 years, but I may find a chance for it in the other world."[10] Her failing eyesight made astronomy impractical, and her health was unreliable, but she refused to give up her habits of study.

In June 1888, she closed a letter to the president of the Vassar board with the sad sentence, "Excuse my writing. It is one of the dark days."[11] She wrote less and less after that, although she did have a photograph of herself taken in front of her new observatory and mailed it to several friends.

On June 28, 1889, Maria Mitchell died at her home in Lynn. The cause of her death is not clear. She was seventy years old, nearly seventy-one. Her death certificate mentioned "brain disease." Shortly after her death, her sister Kate wrote to Mary Whitney, who had succeeded Maria at Vassar, "If only I could wipe out from my memory the last year of her life."[12] Kate chose not to record any of the details she wanted forgotten. Like all of Mitchell's family and friends, her sister was determined to leave a positive picture of Mitchell for posterity.

At Mitchell's funeral on Nantucket, on June 30, 1889, John M. Taylor, then president of Vassar, delivered a eulogy that began the work of softening Mitchell's image. He remarked on her independent spirit and then said, warmly,

> I have sometimes heard this independence dwelt on as if it were the single remarkable trait she possessed and as if the striking speeches and acts to which it sometimes led were alone characteristic of her. But no one who knew Maria Mitchell dwelt most on these. There were great deeps of tenderness and love in the strong nature, a kindliness, a thoughtfulness of others, that beautified her strength and gave balance to her independence. One who has known her kindness to little children, who has watched her little evidences of her thoughtful care for her associates and friends, who has seen her put aside her own long cherished nights that she might make the way of some new and untried officer easier, cannot forget the tenderer side of her character. She was a woman of remarkable strength, but she was a woman withal.[13]

"A woman withal": why did Mitchell's mourners (or Taylor, for that matter) need to be reassured on such a basic point? Did anyone doubt that she was a woman?

Perhaps they did: many late-nineteenth-century obituaries and memoirs of Mitchell included similar lines. The ornithologist Graceanna Lewis, who was careful to soften her portrait of Mitchell by describing her doing laundry, wrote that "Maria Mitchell, like all the distinguished women I have known, was eminently practical, as well as profound, and truly womanly in every fibre of her body."[14]

Educated women in general, and scientific women in particular, had been attacked so often in the last decades of the nineteenth century that defenders of women's education felt it urgently necessary to defend the basic femininity of all distinguished women. Maria Mitchell, who had been singled out as a national heroine before the Civil War, now had to be defended against amorphous questions about her sexuality. Her allies, for the most part, decided to refashion her as a tender-hearted domestic goddess. In 1860, the *Atlantic Monthly* had thought of Mitchell as an American Urania, prefacing Mitchell's essay on Mary Somerville with a poem, "To Urania"— but the 1860s, as Mitchell wrote in that article, were an excellent time for women of genius. By 1889, "Urania" had a very different meaning.

At the beginning of the nineteenth century, people knew Urania as the muse of astronomy; Uranus was the male god of the heavens. The word "Uranian" meant an astronomer, most often a woman astronomer. By the end of the century, Urania had become the muse for writers and sociologists who were trying to define (and often to defend) newly developing forms of homosexual identity. "Uranian" became a medical term for someone with an unconventional sexual sensibility, either a homosexual or person whose psychological identifications ran counter to his or her anatomy: a feminine male or a masculine female. At the middle of the century, calling Mitchell the "American Urania" was a way to celebrate her inspiration of a generation of astronomers. By the time of her death, it would have been a personal insult—an "American Urania" would have been someone who lured unsuspecting followers into the realms of sexual inappropriateness.

Tracing the history of "Urania" and the "Uranians" offers startling insights into the shifting perceptions of women scientists in general, and of Mitchell in particular. In 1891, shortly after Maria Mitchell died and a few weeks after Herman Melville's death, a poem he had written about a woman astronomer was published: "After the Pleasure Party, Lines Traced under an Image of Amor

Threatening." The poem presents sexual self-awareness as a painful reckoning.

Melville's narrator is a woman astronomer named Urania. Her name is rich with significance, since it can refer either to the muse of astronomy or to the identity of the sexual invert. At the beginning of Melville's career, which was roughly contemporary with Mitchell's, Europeans and Americans tended to operate "largely in a realm of undifferentiated sexuality," which is to say that his protagonists Ishmael and Queequeg, like the students at Vassar, shared beds with each other and sometimes traded embraces without necessarily defining themselves in terms of this behavior. In 1845, Margaret Fuller's pronouncement that "there is no wholly masculine man nor feminine woman" had seemed unexceptional. During Melville's and Mitchell's lifetimes, however, sexuality came to be considered "crucial to the construction of the self," and the concept of definable, socially disciplined sexual identity came to the fore.[15] By 1891, a person like Mitchell, who worked in a field that had come to be defined as masculine, could be labeled a Uranian. Melville, whose erotic experiences on the ships and the South Sea islands of his youth had been unconventional and perhaps sometimes homosexual, could also be labeled this way, although neither of them would have thought in such terms thirty years before. Melville's last works, the poems of *Timoleon* and the story of *Billy Budd,* center on dramas of sexual identity. Of these, "After the Pleasure Party" is the most agonized cri de coeur against the inescapability of sexual identity.

"After the Pleasure Party" was published in *Timoleon,* Melville's last, privately printed collection of verse. It is a fairly long poem:160 lines of irregular stanzas five to twelve lines long, capped with a couplet. Urania yearns for sexual love, regrets her choice to focus on the stars rather than on sexual relationships, rails against sexual desire and sexual identity, and moves on to rededicate herself to the goddess Athena, "Transcender" of sexuality. Melville describes his Urania as "Vesta struck with Sappho's smart": that is, a virgin experiencing the agonies of ambiguous sexual desire for an unattainable youth. Urania herself describes sexuality as a "cosmic jest or Anarch blunder."

In 1951, the Melville scholar Leon Howard identified Maria Mitchell as the basis for Melville's Urania.[16] Although few scholars have written about the poem since, those who have generally accept Howard's identification. Most notably, Nathaniel Philbrick has examined the relationship, arguing that although Melville met Mitchell in 1852, when he came to a party at her house above the Pacific Bank, he was moved to write about her only after reading Nathaniel Hawthorne's *Italian Notebooks,* which recounted the events of the winter of 1858 in Rome. Philbrick describes a rueful Melville, who looks back on his passionate friendship with Hawthorne and regrets its stormy end. "In the context of this aching loss," Philbrick writes, "Maria Mitchell's travels with the Hawthornes less than a year after he had been forced to make his own tour of Italy alone...dredg[ed] up a multitude of ambivalent feelings."[17] The sexual ambivalence and regret in Melville's poem is his own as much as Urania's.

The scholars Robert K. Martin and Lawrence Buell both single out "After the Pleasure Party" as one of the most important of Melville's late poems. It is a wholly Melvilleian work: heady, passionate, ambiguous, incredibly well crafted, and a bit of a botch. It is also thoroughly queer. The poem is frank and specific about the pain that comes from being "bound in sex." Melville's Urania struggles with her sexual self-awareness and bitterly mourns her presexual, astronomical nights:

> O terrace chill in Northern air,
> O reaching ranging tube I placed
> Against yon skies, and fable chased
> Till, fool, I hailed for sister there
> Starred Cassiopeia in Golden Chair.
> In dream I throned me, nor I saw
> In cell the idiot crowned with straw.[18]

Urania's mention of her "reaching ranging tube" makes it clear that astronomy is not at all incidental to this discussion: before sex asserted itself, she understood her own identity by means of the telescope, through which she saw the queenly Cassiopeia. Looking constantly through that reaching ranging tube, she was able to look over her sisters on earth, the gibbering madwomen whose sexual frustrations or appetites got them locked up in cells. The telescope acts as both a visual aid and a blindfold: it helps Urania to see herself as powerful, even queenly, while at the same time it blinds her to the perils of sexual subjection. Now she regrets—that is, longs for—her illusions of power and her blindness to her own subjection:

> And yet, ah yet scarce ill I reigned,
> Through self-illusion self-sustained,
> When now—enlightened, undeceived—
> What gain I barrenly bereaved![19]

These are some of the most beautiful lines in the poem. The "self-illusion self-sustained" of the solitary, androgynous astronomer-queen balances perfectly against the sexual pain of the woman "now —enlightened, undeceived." Melville seems to see biology itself as a kind of deception. The poem undeceives its readers about the stability of gender, first by assuming a female persona, then by discussing the frustrations of sexual self-definition, and finally by invoking Urania, the muse of astronomers and sexual inverts.

Once she is undeceived about sexuality, Urania rails against it, calling sex a "cosmic jest or Anarch blunder" that makes "selfhood itself [seem] incomplete." Here Urania alludes to Plato's *Symposium,* in which Aristophanes explains sex as the splitting of whole selves into lonely half selves, doomed to seek their missing other halves. But Plato's story is a humorous one; Melville's version is bitter and painful. For late-nineteenth-century Americans, who were under a great, altogether new pressure to define themselves in terms of "true sex"—as wholly masculine males or feminine females—the moment when "selfhood itself seems incomplete" was a moment of devastating undeception.

At this point the action of the poem moves from the coast of Italy to the city of Rome. "Nerveless" and weak, and longing to escape the "petty hell" of sexuality, Urania considers dedicating herself to the Virgin Mary. She rhymes "grieve" with "believe," and half resolves to "Believe and submit, the veil take on." The veil stands for the convent and a submissive retreat from sexuality, but it also opposes the telescope. A woman who takes the veil obscures her own vision for the sake of protection from the sexually predatory gazes of men, but she also accepts a subordinate position as a protected object of the male gaze. When Urania rejects the veil, she refuses to define herself in terms of others' gazes. She intends to keep looking, even if her newly sexualized gaze might fill her with shameful desires and hopeless frustrations.

Instead of going to the Virgin Mary, Urania turns to the virgin goddess Athena, goddess of wisdom and scholarly inquiry. She prays:

> O self-reliant, strong and free
> Thou in whom power and peace unite,
> Transcender! raise me up to thee,
> Raise me and arm me![20]

The transcendence of sexual identity that Urania requests may be impossible to achieve. Rather than end with her prayer, the poem ultimately comments bleakly that hers is a hopeless appeal, that "Nothing may help or heal" the agonies of sexual awareness. But even if such transcendence is impossible, it's also fascinating. In the figure of the spear-carrying Athena, Urania sees art, history, scientific inquiry, and resistance to sexuality allied. When she asks the goddess to "arm me," she is asking for a spear like Athena's or for a telescope; she wants to look and be looked at. She refuses to be either "sexless" or "bound in sex." Urania hopes passionately for a bold, transcendent androgyny.

"After the Pleasure Party" is virtually unique in its explicit discussion of sexual identity as a blunder, a jest, or a deception. For Melville in the 1890s, it was a matter of fear and threat, and for the woman astronomer he imagined, sexual self-awareness was sheer agony. In part this darkness reflected Melville's own character and circumstances, but we can also read it as a reflection of the great changes in American culture from the early 1860s to the 1890s. Maria Mitchell may not have had the feelings that Melville attributed to her, but she was inevitably caught up in the changing histories of sex and science that he found so momentous.

It is almost impossible to imagine a time when "man" and "woman" were not fixed identities, defined by doctors and professional research scientists. It is equally hard to envision a time before the great divide between science and the humanities, when "scientist" was not a fixed intellectual identity, defined by the same powerful institutions that decide on matters of sexual identity. Perhaps it is hardest to imagine a time when science and sexuality had nothing to do with each other. But when Maria Mitchell was discovering comets, publishing her results in leading journals, and writing profiles of women of genius, and when hundreds of Massachusetts schoolgirls were studying astronomy and botany and reading Mitchell's articles in the *Atlantic,* science seemed a perfectly womanly pursuit.

And then times changed. Doors closed. Following the cultural critics Cynthia Eagle Russett and Gillian Beers, I read the change as a response to the terrifying implications of the new sciences themselves. Because of Copernicanism, Darwinism, and electromagnetism, the world suddenly seemed radically uncertain, and the uncertainties of sex and gender became particularly hard to bear. For everyone, male or female, the scientific thought of the early and middle nineteenth century played merry hell with conventional thinking on identity. Beers tells us, "The great physicist James Clerk Maxwell mused in 1874, that 'we are once more on a pathless sea, starless, windless, and poleless.'"[21] Nineteenth-century science made everything strange and unstable. The whole world, physical and metaphysical, took on the sort of "weird curves" that had long characterized sexual sensibility, and many men responded by trying to reconstruct a lost order—using the very hierarchies of race and sex that so much scientific thought undermined.

And what of nineteenth-century women scientists? Today, when the question of women scientists has once again become a question, it is a good time to remember the women astronomers: the Caroline Herschels and Maria Mitchells, the Uranians. The symbolic history of Urania is a bitter story of exclusion. Women like Maria Mitchell were pathologized, changed from cultural heroes to monsters. Adrienne Rich, in *Planetarium,* called Caroline Herschel a "woman in the shape of a monster." Those who liked Maria Mitchell and supported women as scientists were forced to defend the astronomer's womanhood. After her death there was a concerted effort to domesticate her, to depict her as "a woman withal." Mary King Babbitt, one of her students, wrote an influential biography of Mitchell in 1912 that described Mitchell in her observatory—crocheting. In the 1860s, Vassar alumna Helen Brown described Miss Mitchell's clock room as "a room that to a sensitive girl conveyed a new conception of severe scholarship, that gave her a fresh and inspiring glimpse of the possible attainments of women."[22] By 1912, memories of the room had changed, and Babbitt described it as a quietly welcoming domestic refuge, where "over crochet, and sometimes over tatting, the intellect of our professor of astronomy was wont to unbend itself."[23] Like Nathaniel Hawthorne fifty years before, the defenders of Maria Mitchell in the early twentieth century filled her hands with sewing. Babbitt's brief biography, which John Lankford singles out at the first significant effort to domesticate Mitchell, celebrates the scientist's self-sacrifice and minimizes her achievements. It also emphasizes the idea that Mitchell's astronomical work was much less valuable than her personality. The sketch ends approvingly with a student's comment: "It is not more astronomy that we want; it is more Maria Mitchell." In 1939, when Helen Wright presented the first manuscript of her biography to the Vassar College publishing house, it was rejected with outrage because it focused too much on astronomy, too little on "the woman."[24] For many years, the most important thing that Mitchell's advocates needed to say about her was that she was "truly womanly in every fibre of her body."[25]

These profiles should not be understood as efforts to diminish Mitchell. Instead, they were efforts to defend her in a harsh era when a scientific woman could be libeled "Uranian" and dismissed. But many of Mitchell's most loyal advocates had a hard time recognizing what we would be well advised to remember today: there was a time in this country, not so long ago, when women scientists were not seen as threatening, monstrous, pathological, or deviant. There was a time when Margaret Fuller imagined "the female Newton" with laughter, as one of Nature's "gayer pranks,"[26] and when Maria Mitchell blithely proclaimed to her female students, "Born a woman! . . . what higher destiny could you have? No matter where you are nor what you are, you are a power. Your influence is incalculable."[27] That sense of female possibility is the happy side of Urania's story. The scientific antifeminism of the late nineteenth century belittled, pathologized, and ultimately forgot the Uranian women scientists. It has been left to later generations to remember them.

On a rainy night in late October 2006, a hundred people in formal attire gathered at the Nantucket Inn to celebrate the Maria Mitchell Association's Women in Science Awards. The room was decorated with stars and seashell art projects made by Nantucket schoolchildren. Local science teachers and students mingled with distinguished scientists and wealthy supporters. Carolyn Mabee, the mother of the winner, smiled proudly, but her smile was sad. The 2006 award was presented posthumously to Denice Dee Denton, an engineer and well-known advocate for women in science, who had killed herself in June 2006.

There was much to celebrate in Denton's life. A former colleague, Amy Wendt, told stories about Denton's meteoric rise through the academic hierarchy, from her first appointment as assistant professor in 1987 to that of university chancellor in 2005. Denton had acted as a mentor to Wendt and to scores of other women struggling in hostile disciplines. Denton's friend Nancy Hopkins, a professor of biology at MIT, regaled the crowd with the story of the time, not long before, when the two of them had challenged Lawrence Summers, then president of Harvard, for publicly speculating about differences between men's and women's scientific and mathematical aptitudes.[1] Donna Shalala, former U.S. secretary of health and human services, was not at the dinner, but just a few weeks earlier, she had penned a tribute to Denton in *Beyond Bias and Barriers,* a report on women in science released by the National

Academies of Science and Engineering. The first part of Shalala's dedication could have been written for Maria Mitchell: "Denice Denton was an extraordinarily talented scholar, educational leader, and relentless voice for progress.... She was bigger than life. She opened doors, and stood in them to let others through. She mentored young scholars and students. Her enthusiasm for science was clear and infectious. She was a force—a magnificent force. She pushed the institutions she inhabited to better than they wanted to be."[2]

But although Denice Denton was like Mitchell in many ways, her tragic death made her story very different. One sobering possibility is that the "bias and barriers" that Denton fought in the early twenty-first century may have been even more difficult to surmount than the obstacles facing Mitchell more than 150 years before.

Denton was clinically depressed at the time of her death, and her medical history was a large factor in her depression. But the *Chronicle of Higher Education* reported that a series of "horrible experience[s]" contributed to her depression, starting with "the resistance she often faced as a woman in her field, such as being locked out of her lab" when she was an assistant professor; being so discouraged from doing research that she made the switch to academic administration; and then, as chancellor at the University of Santa Cruz, finding herself the central figure in "an avalanche of bad publicity and outrage" over her pay, in spite of the fact that she was "the lowest-paid chancellor in the system" whose salary was "well below the median for presidents of public research universities." Much of the public outcry stemmed from the fact that the university renovated the chancellor's residence before she arrived, spending $600,000 to upgrade the "modest-looking one story ranch house." Particularly in view of Maria Mitchell's long-standing struggles with the Vassar administration over her housing and compensation, it seems almost eerie that Denton was harassed for her supposedly exorbitant pay and lavish housing while she was the lowest-paid person of her rank in the system, living in one of the smallest houses. There was much public opposition to the simultaneous appointment of her partner, Gretchen Kalonji, as faculty associate to the provost of the University of California system. The attacks were often unrelated to her professional life, the *Chronicle of Higher Education* reported: "The criticism of Ms. Denton was often personal. Students and other critics mocked how Ms. Denton styled her bright shock of strawberry-blond hair, her colorful eyeglasses, and other aspects of her appearance."

It could be argued that Denton had more institutional success than Mitchell, since she rose to be chancellor of a large research university while Mitchell never left her professorship. But even that is debatable, since Denton's friends maintained that she was forced out of research science in part by the hostility she encountered there. And Denton was indisputably the target of sharply personal attacks that made her career far more difficult than Mitchell's. Herman Melville's posthumously published speculations about Maria Mitchell's sexual longings were mild, even sympathetic in comparison to the harsh columns of the *San Francisco Chronicle* attacking Denton and her "lesbian lover."[3]

At the dinner on Nantucket, speakers focused on Denton's achievements rather than her suicide, depression, or public persecution; even so, much of the talk was sobering. One of the episodes that brought Denice Denton onto the national stage was her determination to refute Lawrence Summers's contention that there might be differences in men's and women's "overall IQ, mathematical ability, [and] scientific ability" that were biologically innate, "not plausibly culturally determined."[4] In 2005, under Summers's leadership, Harvard had offered tenured professorships to four women and twenty-eight men. Nancy Hopkins and Denice Denton had used his disparaging comments to launch a national public discussion on women in academic science and engineering.

Hopkins, Denton, and the legions of scholars who joined in their efforts were somewhat successful in refuting Summers's speculation. The Shalala report, which addressed the controversy, drew on twenty-five years' worth of research into sex differences in brain structure and function, hormonal influences on cognitive performance, psychological development, and human evolution in concluding that there was "no clear evidence that men are biologically advantaged in learning and performing mathematics and science."[5] Further, since "men and women do not differ in their average abilities, and because they have now achieved equal success in science through the college level," the report concluded that "there is no sex performance difference for the biological studies and theories to explain."[6] But although girls and women have achieved parity with boys and men up through the college level, there is a disturbing difference between the men's and women's rates of professional success in science: at every stage after college, women leave science in ever increasing numbers. Fewer women enter graduate programs in the sciences; even fewer take jobs as researchers; and very few indeed become tenured senior professors in scientific fields. *Beyond Bias and Barriers* reports that "at the top research institutions only 15.4% of the full professors in the social and behavioral sciences and 14.8% in the life sciences are women—and these are the only fields in science and engineering where the proportion of women reaches into the double digits."[7]

The commission concluded, "It is not lack of talent, but unintentional biases and outmoded institutional structures that are hindering the access and advancement of women."[8] Universities, they wrote, must move to "an inclusive model with provisions for equitable and unbiased evaluation of accomplishment, equitable allocations of support and resources, pay equity, and gender-equal family leave policies."

These recommendations are strikingly similar to the goals Maria Mitchell articulated in the 1860s and 1870s. Throughout her career, Mitchell stressed the importance of women being hired for professional jobs, paid fairly, and encouraged rather than discouraged. Regarding the proportion of women faculty, she once asked, "Do you know of any case in which a boy's college has offered a professorship to a woman? Until you do, it is absurd to say that the highest learning is within the reach of women."[9] Regarding pay, one of Mitchell's students, Anna Brackett, commented, "The indignant protest" with which she "called for an equal salary, was not a personal affair. She flamed out on behalf of all women, and of abstract justice."[10] As for efforts to encourage women, the *New York Times* reported in 1881, "The question of equality of the sexes is one that does not naturally disturb such a woman. She merely says if women are regarded as equals, in mental capacity, they should have equal advantages, and if considered inferior, they should be given better chances."[11]

Mitchell's 1875 remark that "science needs women" could have served as an epigraph for the 2007 Shalala report, which concluded,

"The United States can no longer afford the underperformance of our academic institutions in attracting the best and the brightest minds to the science and engineering enterprise . . . It is essential that our academic institutions promote the educational and professional success of all people without regard for sex, race or ethnicity."[12] In spite of these positive goals, the statistics are bleak. In many respects, women scientists have moved backward since the 1860s. When Vassar opened its doors in 1865, the sole woman professor on the faculty was a scientist; the only other woman of comparable rank was the college physician. At that time, the stereotypes still worked the other way; an 1866 book in Mitchell's own library on the higher education of women explained, "In physical science, astronomy and botany are considered the ladies' department."[13] Mitchell's niece Rebecca Coffin, who enrolled at Vassar hoping to study fine arts, knew more successful women scientists than successful women painters. Tracing the path of American women in science sometimes seems akin to charting the orbit of Venus. In the early nineteenth century the idea of educating women was progressive, but for those who were to be educated, science was seen as the most appropriate subject. By the beginning of the twentieth century, women were decisively steered away from science, but education for women was gaining wide acceptance. It can be hard to gauge progress and account for women's reversals amid these cultural contradictions.

For nineteen years Maria Mitchell worked as the computer of Venus, using complex logarithms to predict when the planet would progress and when it would retrogress. Around the time she resigned her position at the *Nautical Almanac,* in 1868, the climate began to worsen for scientific women: a great retrogression had begun. The opportunities that had helped Mitchell and her Nantucket sisters to thrive became increasingly rare.

One of the strangest aspects of the reversals for scientific women was that the atmosphere of encouragement that had fostered Mitchell became unimaginable to her students. Looking back through their own discouraging experiences, they imagined that the young Mitchell had fought the same prejudices. In 1879, a profile of Mitchell in *Godey's Lady's Book* framed her story with exaggerated accounts of the obstacles girls had faced in the 1830s and 1840s:

"Properly to appreciate Miss Mitchell's wonderful attainments," the author claimed, "we must remember that 40 years ago for a woman to distinguish herself in scientific knowledge was a very different matter from what it is today. We read of... procuring books by stealth and studying them secretly at night while others slept to conceal the unfeminine proclivities.... In those days, there were no colleges for women, and no fraternity of monks could have closed their gates more sternly against an intruder in feminine apparel than did the professions generally. A woman's examination of the stars was supposed to take place only in a poetic mood and under poetic circumstances." After setting the scene this way, the writer acknowledges that Mitchell was not excluded from astronomy: "It is pleasant to know that [Mitchell] was not obliged to knock fruitlessly at the door of any institution of learning or to wander disheartened on a woman's quest for instruction under difficulties, but that in the privacy of a cultured home, and under her father's influence, she found all that she needed."[14] It is striking that this 1879 account imagines great difficulties for scientifically inclined girls in the beginning of the nineteenth century, even though the writer must admit that Mitchell's family and community encouraged her scientific aspirations. By the late nineteenth century, it was already hard to envision an atmosphere in which girls were encouraged to study science. By the middle of the twentieth century, Vassar historians tended to deny that Mitchell had ever received any encouragement at all. In 1940, Vassar alumna Martha Hillard Macleish summed up the prevailing accounts of Mitchell's past by explaining that "a woman astronomer was practically unknown in those days."[15]

These claims were deeply inaccurate. The fact is that Mitchell studied astronomy in a time when girls were more encouraged to study science than boys were. It is true that there were no colleges for women during Mitchell's girlhood, but her chroniclers overlooked the fact that there were no organized astronomy courses in any college at that time. Whatever her enormous accomplishments, Mitchell had not triumphed over circumstances; to the contrary, she had been born into the most propitious possible situation for a young astronomer. Perhaps her most significant advantage was that her father was a respected astronomer, but being raised in Massachusetts, which had a strong tradition of common schools for both sexes and was developing sophisticated astronomy courses for its girls' high schools, was also an advantage. Nantucket's Quaker traditions ensured that Maria's mind would be cultivated regardless of her sex, while its reliance on seafaring meant that the surrounding community shared her interest in the stars, at least insofar as they were important to navigation.

When Mitchell burst onto the national stage with her discovery of the comet, many people expressed surprise at her great intellectual abilities, but few commented on her proclivity for science as opposed to other, softer endeavors. The liberal nineteenth-century journalist Thomas Wentworth Higginson wrote in a Massachusetts paper, "Hundreds of women have in recent years distinguished themselves in scientific pursuits, but we are always interested in pioneers."[16] Similarly, *Scientific American* contextualized Mitchell's achievements with the accomplishments of other women interested in science: "Science has its followers among the gentler sex, but among them the name of Maria Mitchell stands out clear and conspicuous, like an evening star in the heavens she loved so well to study."[17] In both of these nineteenth-century essays, Mitchell is described as merely the most accomplished among a host of scientific women.

But as times changed, commenters became more and more likely to focus on Mitchell's strangeness. In 1872, members of the Association for the Advancement of Women applauded wildly when Mitchell declared, "Science needs women." By 1891, the audience at the National Council of Women of the United States applauded just as enthusiastically when Harriet Stanton Blatch proclaimed that "to rear an astronomer is perchance a higher labor than to discover a comet."[18] Women astronomers, along with other women scientists, had spiraled backward.

In spite of the increasing barriers women would face in the late nineteenth and early twentieth centuries, some exceptional women continued to make forward progress. Marie Curie won Nobel Prizes for chemistry and physics in 1903 and 1911. Like Mitchell, she endured much adverse publicity regarding her sexual inappropriateness—while Mitchell had become, by the 1940s, a mythical "astronomical nun,"[19] Curie was publicly bashed for being a monstrous mother. Both women were reported as exceptions or anomalies in a way that used their achievements to emphasize the unlikelihood of women acting as scientists at all.

Eventually, the myth reshaped reality. In 1959, C. P. Snow casually remarked in *The Two Cultures and the Scientific Revolution*—a landmark in the history of science—"We don't in reality regard women as suitable for scientific careers."[20] Shortly thereafter, in 1962, Thomas Kuhn published *The Structure of Scientific Revolutions,* in which, like Snow, he consistently described all scientists as "men." Although women scientists were relatively rare in the 1960s, they were certainly known to exist, no matter the definition of scientist: Maria Mitchell had never quite been forgotten, and Marie Curie had won Nobel Prizes. Women in white coats stood at every laboratory bench in every college and university research lab around the world, but they were classed as assistants or technicians, not scientists. The few women who achieved the status of scientists were imagined to have abandoned their own femininity.

Even at Vassar College, the idea of choosing a career as a scientist became anathema by the middle of the twentieth century. When Betty Friedan visited Vassar in 1959, the tide had turned completely. "What courses are people excited about now?" Friedan asked a "blonde senior in a cap and gown."

> Nuclear physics, maybe? Modern Art? The civilizations of Africa? Looking at me as if I were some prehistoric dinosaur she said:
>
> "Girls don't get excited about things like that any more. We don't want careers. Our parents expect us to go to college. Everybody goes. You're a social outcast at home if you don't. But a girl who got serious about anything she studied—like wanting to go on and do research—would be peculiar—unfeminine. I guess everybody wants to graduate with a diamond ring on her finger. That's the important thing."[21]

• • •

Throughout the twentieth century, cultural denial of the existence of women scientists took two basic shapes. On the one hand, women were discouraged from studying science, denied entry to graduate programs, made ineligible for fellowships or grants, steered away from research, and excluded from professional organizations. At the same time, the women who persisted in working in science were overlooked, and their careers as science teachers, science writers, and laboratory and field researchers were considered unprofessional. Many of the women who devoted their lives to scientific work were carefully defined as nonscientists; the identity of the scientist was reserved for the professionals at the top of the institutional hierarchy. As a kicker, the few women who did manage to gain professional and institutional status soon discovered, to their own surprise, that they were no longer seen as women. The association between science and masculinity had grown so strong that women's successes in science were often belittled as failures of femininity.

Many courageous women scientists and feminist scholars battled these biases throughout the twentieth century. These women made some gains, but the climate for women in the sciences at the beginning of the twenty-first century is still much harsher than the corresponding climate for women in politics, law, business, medicine, journalism, or the humanities. The sciences remain the most rigidly masculine professional fields in the United States.

In January 2005, Lawrence Summers blundered into a minefield with a speech that attempted to explain women's exclusion from the sciences in terms of their innately unscientific qualities. Denice Denton and Nancy Hopkins heard his speech and called the newspapers, and the next two years were filled with vociferous debates over women in the sciences. The outrage against both Denton and Summers was public and prolonged. Summers survived it, as Denton did not, but eventually he was asked to resign from the presidency of Harvard. In fall 2007, Summers was replaced by Drew Gilpin Faust, the university's first female president. It remains to be seen whether Faust, a historian by training, can provide the desperately needed leadership that will welcome women into the sciences at Harvard, but of course Harvard is only a small part of the picture. All American educational institutions need to do more to welcome women into professional science, in part to honor basic principles of fairness, but in larger part because the sciences suffer when talented people are excluded from them.

As the tragic story of Denice Denton's withdrawal from the fight indicates, the battle for women in science can be deeply demoralizing. Maria Mitchell's story offers hope to those who carry this torch. The fact that American girls and women were expressly encouraged to study science from around the time of the American Revolution until the late nineteenth century means that such an age can surely come again. Women scientists can build new institutions that make science more inclusive and more successful. Maria Mitchell's life and career show the possibilities for women who are given opportunity, as well as some of the constraints that have limited women scientists thus far. Mitchell asked her students at Vassar, "Does anyone suppose that any woman in all the ages has had a fair chance to show what she could do in science?...Until able women have given their lives to investigation, it is idle to discuss the question of their capacity for original work."[22]

ACKNOWLEDGMENTS

Simmons College has been an ideal place to explore the life and legacy of a pioneer in women's education. The extraordinary women students at Simmons are a constant inspiration, and the brave men who study among them in our graduate programs are equally awesome. In the English Department, Kristina Aikens, Sheldon George, David Gullette, Jane Kokernak, Kelly Hager, Kendra Muelling Carter, Cathie Mercier, Douglas Perry, Afaa Michael Weaver, and Richard Wollman are teachers, scholars, poets, and witty colleagues who make my workplace intellectually challenging and wholly satisfying. I miss the ones that got away: Catherine Allgor, Burlin Barr, Eileen Cleere, and Keith Gorman. I offer particular thanks to Pamela Bromberg and Lowry Pei, who have served as chairs of the department and as generous and thoughtful mentors to me. I'm also grateful to many others at Simmons, including: Janet Chumley, Wanda Torres Gregory, Mary Jane Treacy, Velda Goldberg, Jyoti Puri, and Jill Taylor. In Gender/Cultural Studies, Denise Oberdan and Meg Killian offered comradeship along with administrative derring-do, and in the English Department, Rachel Ruggles was a beacon of clear thinking and radiant efficiency who regularly made impossibilities possible.

A research appointment in Studies of Women, Gender, and Sexuality at Harvard has offered me access to the light-filled atrium hidden in the heart of Widener Library, and to the miles of books and acres of intellectual buzz that surround it. I am particularly grateful to Robin Bernstein for making me welcome at Harvard. Meanwhile, in the hills far from Boston, at Dartmouth College, Ivy Schweitzer and Tom Luxon are friends and mentors whom I rely on and admire. This book was written in a tiny study tucked into the wings of Baker Library. Dartmouth's generosity in providing a visitor with a room of her own, a vast network of resources, and a staff of crack librarians has shown me what an ideal intellectual community might look like.

Many generous fellowships supported work on this project. An NEH We the People Grant supported the final year of writing. A Simmons College Fund for Research grant helped me to get started and another helped me tie up some loose ends. With help from Simmons, I was able to visit Nantucket in the early stages of the project. Later, I went to Vassar College, where the College historian, Betty Daniels, was a warm hostess. Her intrepid guidance around the (soon to be renovated) Vassar College Observatory brought me close to Mitchell's spirit, and gave me great appreciation for Vassar's generations of women and men. Dean Rogers was a friendly guide through the Vassar College Archives.

A Caleb Loring, Jr., Fellowship from the Boston Athenaeum in 2005 helped me to understand the impact of the Civil War on American education and science, while an E. Geoffrey and Elizabeth Thayer Verney fellowship from the Nantucket Historical Association in 2006 gave me the opportunity to live on Main Street for a few weeks, and to immerse myself in the salty air of Mitchell's hometown. William Tramposch, Ben Simons, and Elizabeth Oldham helped to make my time on Nantucket pleasant and productive. The Nantucket Atheneum was also a very useful and welcoming library. Charlotte Maison helped to get me started on my research, and Sharon Carlee helped me to track down some of the most important finishing details.

The Maria Mitchell Association on Nantucket maintains her birthplace and her archives. Its programs welcome children and college students who are interested in the natural world to study many sciences from marine biology to astronomy. When I started work on the project, Patty Hanley welcomed me into the archives. Later, Amy Hunt England offered me a bed in a Mitchell Association house, Jascin Finger helped me in the archives, and Janet Schulte was a challenging interlocutor, an impressive leader, and best of all, a friend.

My agent, Giles Anderson, helped me figure out how to put this book together, and find a press for it. At Beacon, Christine Cipriani did Herculean editorial labor. Tracy Roe copyedited the manuscript with elegance and insight, while Lisa Sacks was generous with pencils and attention, and Sarah Gillis carefully steered the book through production. Amy Caldwell added the finishing touches. I am grateful to all of them.

I'm also thankful for American Studies friends: Karen Sànchez-Eppler, Laura Wexler, Priscilla Wald, Dana Nelson, Greg Jackson, Russ Castronovo, Rachel Adams, Caroline Levander, Mary Kelley, Michael Elliott, and particularly Lloyd Pratt, who introduced me to Maria Mitchell when I was searching for a nineteenth-century woman "with good eyes." My NEASA buddies, Lisa Macfarlane, Mary Battenfeld, Eve Raimon, Donna Cassidy, and the amazing members of the Council have helped me to feel part of a local community of scholars whom I value very much.

I have been very fortunate to be part of a writing group that grew out of NEASA. Sarah Luria, Laura Saltz, Augusta Rohrbach, and Betsy Klimasmith shared many of the stages of this book with me, and their friendship, support, and incredible collective intelligence has been a mainstay and a source of great happiness for me. I am particularly grateful to Betsy Klimasmith, who read every word of this manuscript when there were far too many of them, and whose support and insightful suggestions helped me to push on.

My father, Richard M. Bergland, and my father-in-law, Truls Brinck-Johnsen, are both scientists who have always made it clear to me that they respected women's intelligence and shared the belief that science needs women. My mother-in-law, Kari Brinck-Johnsen, was one of the "girls in the lab," at Dartmouth College, and the paradoxes of her position gave me great insight into women in science.

When I write about the joys of a warm, intelligent, encouraging family, I am writing from experience. I wrote the proposal for this book in the garret at my mother's house in the Berkshires, and I moved into her sweet little chicken coop when it was time to finish the manuscript. My mother, Mary Jo Litchard, welcomed the fragile volumes of nineteenth-century poetry, letters, and memoirs that I stacked around her house. Meanwhile, my sister and brother both read the manuscript with loving care. Their interest and encouragement have meant a great deal to me. My sister, Sandy Bergland, has supported this project, sometimes with cold hard cash, sometimes with bags of dog food, brave offers to transport the cat to the vet, or panting conversations as we run together. I thank her for her encouragement and help. My brother, Christopher Bergland, wrote a book of his own while I was working on this one—sharing the ups and downs of the writing process with him has been a source of comfort and strength. My niece, Helene Margel, also lived with us during the year I was writing this book. She is a brave girl, and I am thankful for her tender heart, her thoughtfulness, and her giant smile.

Kim Brinck-Johnsen and I have lived and worked together for many years, and his partnership sustains me. The winter that I finished this manuscript we were lucky enough to have little offices a few doors apart from each other. Our partnership offers us opportunities for solitude and warm companionship in equal measure. It is a rare good fortune to marry someone who is gentle in spirit, good to the core, and unfailingly generous. I am thankful that I share my life with a man who is all that, and who is also a great cook, an intelligent critic, and a master of mixology. In the context of this book, I am particularly grateful that Kim and Stjerne walked along the river with me every morning on the way to the library, and that Kim slogged through every word of the manuscript, challenging and reassuring with intelligent precision, and finally managing to track down Ida Russell, woman of mystery.

My daughter, Annelise Bergland Brinck-Johnsen, was five years old when we went on the first Maria Mitchell research trip to Nantucket. She pedaled furiously over the cobblestones of Main Street on her trail-a-bike and colored diligently at a small low table at the Maria Mitchell Association library while I read Mitchell's notebooks. Those mornings in Nantucket were a great pleasure for both of us, and I was amazed that she could be so patient and happy in the library when she didn't yet know how to read. I am even more amazed that in just a few years she has become one of my best readers and most energetic research assistants. She is an intrepid finder of books in the recesses of the Boston Athenaeum and Baker Library, and her insights startle and delight. Annelise has grown up with Maria Mitchell, and I think that Mitchell has been an encouraging presence in her life. I know that this book is much better because I wrote it in the company (and with the help) of a kindhearted and ferociously intelligent girl.

Introduction

1. Lilla Barnard, "Reminiscence of Maria Mitchell," *Woman's Journal* (MMP 55); cited in Henry Albers, ed., *Maria Mitchell: A Life in Journals and Letters* (Clinton Corners, NY: College Avenue Press, 2001), 21.

2. Ibid.

3. Helen Wright, *Sweeper in the Sky: The Life of Maria Mitchell* (1949; repr., Clinton Corners, NY: College Avenue Press, 1997), 87.

4. David Alan Grier, *When Computers Were Human* (Princeton, NJ: Princeton University Press, 2005), 60–63.

5. Ibid., 63.

6. Sally Gregory Kohlstedt, "Maria Mitchell: The Advancement of Women in Science," *New England Quarterly* 51, no. 1 (1978): 57.

7. Ibid.

8. Sandra Harding, *The Science Question in Feminism* (Ithaca, NY: Cornell University Press, 1986), 31.

9. Edward H. Clarke, *Sex in Education, or, a Fair Chance for Girls* (Boston: James R. Osgood, 1874), 115.

10. William Whewell, "Review of 'On the Connexion of the Physical Sciences,'" *Quarterly Review* 51 (1834): 54.

11. Kathryn A. Neeley, *Mary Somerville: Science, Illumination and the Female Mind* (Cambridge: Cambridge University Press, 2001), 3.

12. Lucretia Mott, "Discourse on Woman," 1849;.

13. John H. Raymond, "The Demand of the Age for the Liberal Education of Women and How It Should Be Met," in James Orton, ed., *The Liberal Education of Women: The Demand and the Method* (New York: Garland Publishing, 1986, facsimile reprint of 1873 text), 49–50.

14. Ibid.

Chapter 1: Urania's Island

1. Julia Ward Howe, "Maria Mitchell," *Our Famous Women* (Hartford, CT: A. D. Worthington, 1884), 454.

2. William Mitchell, "Autobiography of William Mitchell," in Helen Wright, ed., *Historic Nantucket* 30, no. 3 (1983): 19.

3. Phebe Mitchell Kendall, ed., *Maria Mitchell: Life, Letters, and Journals* (1896; repr., Freeport, NY: Books for Libraries Press, 1971), 15.

4. Howe, *Our Famous Women,* 441.

5. Kendall, ed., *Life, Letters,* 8.

6. Anne Mitchell, "Reminiscences of the Family Life," *Friend's Intelligencer,* December 21, 1889.

7. Henry Albers, ed., *Maria Mitchell: A Life in Journals and Letters,* (Clinton Corners, NY: College Avenue Press, 2001), 58.

8. Anne Mitchell Macy, "Astronomical Garret," Maria Mitchell Association, n.p., n.d.

9. Samuel Haynes Jenks, *Nantucket Inquirer,* October 18, 1828.

10. Howe, *Our Famous Women,* 444.

11. Albers, ed., *Maria Mitchell,* 156.

12. Kendall, ed., *Life, Letters,* 9.

13. William Mitchell, "Autobiography," in *Historic Nantucket,* 16.

14. Ibid., 42–43.

15. Henry Mitchell, "Biographical Notice of Maria Mitchell," *Proceedings of the American Academy of Arts and Sciences* 25 (1889), 331.

16. Ibid.

17. Ibid.

18. Howe, *Our Famous Women,* 444.

19. Franklin to Mary "Polly" Stevenson Hewson, May 17, 1760; www.franklin papers.org/franklin/framedNames.jsp?ssn=001–72–0019.

20. Ibid., May 1, 1760.

21. Will Gardner, *The Clock That Talks and What It Tells* (Nantucket, MA: Nantucket Whaling Museum, 1954), 27.

22. Lisa Norling, *Captain Ahab Had a Wife: New England Women and the Whalefishery, 1720–1870* (Chapel Hill: University of North Carolina Press, 2006), 111.

23. This book is in Special Collections at Harvard University. The book was passed from generation to generation until 1946, when Folger Coleman's great-granddaughter Laura H. D. Saunderson sold it to a collector (Philip Hofer), who gave the book to the Houghton Library at Harvard.

24. The poem is in the commonplace book "Un Recueil." A version of the poem that was published in the *Cleveland Daily Advertiser* in 1837 is in the archives at the Nantucket Historical Association (NHA), along with a handwritten

letter from Phebe Folger Coleman's daughter and an alternative (more correct) version of the poem—the newspaper editors had changed shared "books" to shared "looks," and made a few other changes worthy of comment; NHA Collection 353 (Envelope).

25. Gardner, *The Clock That Talks,* 82.

26. National Historical Association, Collection 80, Book 1.

Chapter 2: Nantucket Athena

1. Sally Gregory Kohlstedt, "Parlors, Primers, and Public Schooling: Education for Science in Nineteenth-Century America," in Barbara Laslett and others, eds., *Gender and Scientific Authority* (Chicago: University of Chicago Press, 1996), 177.

2. Will Gardner, *The Clock That Talks and What It Tells* (Nantucket, MA: Nantucket Whaling Museum, 1954), 81.

3. Ibid., 15.

4. Ibid., 1.

5. Carl F. Kaestle, *Pillars of the Republic 1780–1860* (New York: Hill and Wang, 1983), 124.

6. Alexander Starbuck, *The History of Nantucket, County, Island, and Town, Including Genealogies of the First Settlers* (Boston: C. E. Goodspeed, 1924), 600.

7. Ibid., 602.

8. Ibid., 613.

9. William Mitchell, "Autobiography," in Helen Wright, ed., *Historic Nantucket* 9.

10. Ibid.

11. Helen Wright, *Sweeper in the Sky: The Life of Maria Mitchell* (1949; repr., Clinton Corners, NY: College Avenue Press, 1997), 32.

12. Arthur M. Kennedy, "The Athenaeum: Some Account of Its History from 1814 to 1850," *Transactions of the American Philosophical Society,* New Series 43, no. 1 (1953): 260–61.

13. Nantucket Atheneum, proprietors' records, 1853.

14. Nantucket Historical Association (NHA), Joy Family Papers, folder 12.

15. Ibid.

16. Ibid.

17. Henry Albers, ed., *Maria Mitchell: A Life in Journals and Letters* (Clinton Corners, NY: College Avenue Press, 2001), 22.

18. Ibid., 23.

19. Wright, *Sweeper in the Sky,* 37–38.

20. NHA Collection 80, Book 1.

21. Maria Mitchell Association, Maria Mitchell Papers (MMP), vol. 16; Dec. 3, 1845.

22. Ibid., 1853.

23. Wai-Chee Dimock and Priscilla Wald, "Preface: Literature and Science: Cultural Forms, Conceptual Exchanges," *American Literature* 74, no. 4 (2002): 705–14.

24. Laura Otis, ed., *Literature and Science in the Nineteenth Century* (New York: Oxford World's Classics, 2002), 79.

25. Phebe Mitchell Kendall, ed., *Maria Mitchell: Life, Letters, and Journals* (1896; repr., Freeport, NY: Books for Libraries Press, 1971), 186.

26. Miriam Levin, *Defining Women's Scientific Enterprise: Mount Holyoke Faculty and the Rise of American Science* (Lebanon, NH: University Press of New England, 2005), 3.

27. Nina Baym, *American Women of Letters and the Nineteenth-Century Sciences* (New Brunswick, NJ: Rutgers University Press, 2002), 130.

28. Emily Dickinson, *The Complete Poems of Emily Dickinson,* Thomas A. Johnson, ed. (Boston: Back Bay Books, 1976), 578.

29. Brad Ricca, "Emily Dickinson: Learn'd Astronomer," *Emily Dickinson Journal* 9, no. 2 (2000): 96–108; James Guthrie, *Emily Dickinson's Vision: Illness and Identity in Her Poetry* (Gainesville: University Press of Florida, 1998), 33; Richard B. Sewall, *The Life of Emily Dickinson* (Cambridge, MA: Harvard University Press, 1994), 354.

30. Kendall, ed., *Life, Letters,* 46.

31. Kennedy, "The Athenaeum," 260–61.

Chapter 3: The Sexes of Science

1. Margaret Alic, *Hypatia's Heritage* (Boston: Beacon Press, 1986), 126.

2. Mrs. John (Mary) Herschel, *Memoirs and Correspondence of Caroline Herschel* (New York: Appleton, 1876), 167.

3. John Lankford, *American Astronomy: Community, Careers, and Power, 1859–1940* (Chicago: University of Chicago Press, 1997), 341.

4. Ibid.

5. Jan Golinski, "The Care of the Self and the Masculine Birth of Science," *History of Science* 40, pt. 2, no. 128 (June 2002): 125–45.

6. Patricia Fara, *Pandora's Breeches: Women, Science and Power in the Enlightenment* (London: Pimlico, 2004), 55–73.

7. Sally Gregory Kohlstedt, "Parlors, Primers, and Public Schooling: Education for Science in Nineteenth-Century America," *Isis* 80, no. 3 (Sept. 1990), 430–31.

8. Londa Schiebinger, *The Mind Has No Sex? Women in the Origins of Modern Science* (Cambridge, MA: Harvard University Press, 1989), 145.

9. Fara, *Pandora's Breeches,* 18.

10. David Cahan, "Looking at Nineteenth-Century Science," in his *From Natural*

Philosophy to the Sciences: Writing the History of Nineteenth-Century Science (Chicago: University of Chicago Press, 2003), 4.

11. Michel Foucault, *The History of Sexuality: An Introduction* (New York: Vintage, 1980).

12. Thomas Laqueur, *Making Sex: Body and Gender from the Greeks to Freud* (Cambridge, MA: Harvard University Press, 1992), 20.

13. Helen Lefkowitz Horowitz, *Rereading Sex: Battles over Sexual Knowledge and Suppression in Nineteenth-Century America* (New York: Alfred A. Knopf, 2002), 30.

Chapter 4: Miss Mitchell's Comet

1. Henry Albers, ed., *Maria Mitchell: A Life in Journals and Letters* (Clinton Corners, NY: College Avenue Press, 2001), 35.

2. William Mitchell, "Autobiography of William Mitchell," in *Historic Nantucket* 30, no. 4 (1983): 42–43.

3. Julia Ward Howe, "Maria Mitchell," *Our Famous Women* (Hartford: A. D. Worthington, 1884), 445.

4. Anne Mitchell, "Reminiscences of the Family Life," *Friend's Intelligencer,* December 21, 1889.

5. Albers, ed., *Maria Mitchell,* 29.

6. Ibid., 27.

7. Ibid.

8. Phebe Mitchell Kendall, ed., *Maria Mitchell: Life, Letters, and Journals* (1896; repr., Freeport, NY: Books for Libraries Press, 1971), 21.

9. Albers, ed., *Maria Mitchell,* 30.

10. Plutarch, *Plutarch's Lives,* trans. John Dryden, rev. A. H. Clough (Boston: Little, Brown, 1885), 327.

11. William Shakespeare, *Julius Caesar,* act 2, scene 2.

12. John Milton, *Paradise Lost,* 2:710–11.

13. Mary Wollstonecraft Shelley, *Life and Letters of Mary Wollstonecraft Shelley,* ed. Mrs. Julian Marshall (London: R. Bentley, 1889), 59.

14. Patricia Fara, *Pandora's Breeches: Women, Science, and Power in the Enlightenment* (London: Pimlico, 2004), 160.

15. John Herschel, *A Treatise on Astronomy* (Philadelphia: Lea and Blanchard, 1851), 340.

16. Ibid., 373.

17. Robert Grant, *History of Physical Astronomy: From the Earliest Ages to the Middle of the Nineteenth Century* (London: R. Baldwin, 1852), 138.

18. P. B. Boyce et al., "Maria Mitchell's Comet," Smithsonian/NASA Astrophysics Data System, http://adsabs.harvard.edu/abs/1997AAS...191.3801B.

19. Dorritt Hoffleit, "Comets over Nantucket," Maria Mitchell Association, 1983, 12.

20. Bessie Zaban Jones and Lyle Gifford Boyd, *The Harvard College Observatory: The First Four Directorships, 1839–1919* (Cambridge, MA: Belknap Press of Harvard University Press, 1971), 56, 110.

21. Hoffleit, "Comets," 14.

22. Ibid.

23. I. Bernard Cohen, ed., *Aspects of Astronomy in America in the Nineteenth Century* (New York: Arno, 1980), 61–62.

24. Steven J. Dick, *Sky and Ocean Joined: The U.S. Naval Observatory, 1830–2000* (New York: Cambridge University Press, 2003), 57.

25. Albers, ed., *Maria Mitchell,* 31.

26. Dick, *Sky and Ocean,* 119.

27. Edouard Stackpole, foreword to *Two Steps Down,* by Alice Albertson Shurrocks (Nantucket, MA: Inquirer and Mirror Press, 1953).

Chapter 5: "A Center of Rude Eyes and Tongues"

1. Henry Albers, ed., *Maria Mitchell: A Life in Journals and Letters* (Clinton Corners, NY: College Avenue Press, 2001), 43–44.

2. Ibid., 64.

3. Ibid., 62.

4. Ibid., 64.

5. Ibid., 50.

6. Margaret Fuller, *Woman in the Nineteenth Century* (London: H. G. Clarke, 1845), 107.

7. Ibid., 108–9.

8. Ibid., 30, 166.

9. Albers, ed., *Maria Mitchell,* 54.

10. Ibid., 55.

11. John Lankford, *American Astronomy: Community, Careers, and Power, 1859–1940* (Chicago: University of Chicago Press, 1997), 303–4.

12. Hector St. John de Crèvecoeur, *Letters from an American Farmer* (New York: Fox and Duffield, 1904), 205.

13. Ibid., 208.

14. Ibid., 210–11.

15. Nathaniel Philbrick, *Away Off Shore: Nantucket Island and Its People, 1602–1890* (Nantucket, MA: Mill Hill Press, 1994), 257.

16. Nathaniel Philbrick, "The Nantucket Sequence in Crevecoeur's Letters from an American Farmer," *New England Quarterly* 64, no. 3 (September 1991): 420.

17. Lisa Norling, *Captain Ahab Had A Wife: New England Women and the Whalefishery, 1720–1870* (Chapel Hill: University of North Carolina Press, 2006), 216.

18. Ibid., 217.

19. Fuller, *Woman in the Nineteenth Century,* 103–4.

20. Thomas J. Brown, *Dorothea Dix, New England Reformer* (Cambridge, MA: Harvard University Press, 1998), 216.

21. Albers, ed., *Maria Mitchell*, 42.

22. Brown, *Dorothea Dix*, 125.

23. Albers, ed., *Maria Mitchell*, 46.

24. Ibid., 43.

25. Fuller, *Woman in the Nineteenth Century*, 104.

26. Caroline Healey Dall, *Margaret and Her Friends* (Boston: Roberts Brothers, 1895), 19.

27. Nathaniel Hawthorne, *Centenary Edition of the Works of Nathaniel Hawthorne*, vol. 15, *The Letters, 1813–1843*, Thomas Woodson and others, eds. (Columbus: Ohio State University Press, 1987), 597.

28. Ibid.

29. Theodore Maynard, *Orestes Brownson: Yankee, Radical, Catholic* (New York: Hafner Press, 1971), 115.

30. John Farina, ed., *Hecker Studies: Essays on the Thought of Isaac Hecker* (New York: Paulist Press, 1983), 158.

31. Caroline Ticknor, *Poe's Helen* (New York: Charles Scribner's Sons, 1916), 201.

32. Kenneth Silverman, *Edgar A. Poe: Mournful and Never-Ending Remembrance* (New York: HarperCollins, 1991), 348.

33. Letter, January 7, 1846 (in Whitman papers at Brown University).

34. Albers, ed., *Maria Mitchell*, 61.

35. Maria Mitchell Association, November 14, 1854.

36. Albers, ed., *Maria Mitchell*, 50.

37. Maria Mitchell Association, Maria Mitchell Papers (MMP), vol. 16; November 14, 1854.

38. Albers, ed., *Maria Mitchell*, 51.

39. Ibid., 53.

40. Ibid., 58.

41. MMP, Volume 16.

42. Ibid.

43. "Female Astronomers," *Emerson's United States Magazine* 4, no. 1 (January 1857): 89.

44. Ella Rodman Church, "Maria Mitchell," *Godey's Lady's Book*, November 1879.

Chapter 6: The Shoulders of Giants

1. Maria Mitchell Association, Maria Mitchell Papers (MMP), vol. 16; December 3, 1845.

2. Margaret Fuller, *At Home and Abroad* (Boston: Crosby, Nichols and Company, 1856), 250.

3. Henry Albers, ed., *Maria Mitchell: A Life in Journals and Letters* (Clinton Corners, NY: College Avenue Press, 2001), 81.

4. Herman Melville, *Redburn: His First Voyage* (New York: Harper and Brothers, 1850), 208, 209.

5. Albers, ed., *Maria Mitchell,* 89.

6. Phebe Mitchell Kendall, ed., *Maria Mitchell: Life, Letters, and Journals* (1896; repr., Freeport, NY: Books for Libraries Press, 1971), 86.

7. Ibid., 87.

8. Ibid., 87, 88.

9. Ibid., 96.

10. Ibid., 86.

11. Albers, ed., *Maria Mitchell,* 89.

12. T. Walter Herbert, *Dearest Beloved* (Berkeley: University of California Press, 1993), 263.

13. Kendall, ed., *Life, Letters,* 94.

14. Charles Dickens, *Bleak House* (1854; repr., New York: Penguin, 2006), 1.

15. Kendall, ed., *Life, Letters,* 104.

16. Albers, ed., *Maria Mitchell,* 91.

17. Kendall, ed., *Life, Letters,* 99.

18. Ibid., 96.

19. Ibid., 116.

20. Virginia Woolf, *A Room of One's Own* (New York: Oxford World's Classics, 1998), 10.

21. Kendall, ed., *Life, Letters,* 112, 119.

22. Ibid., 121.

23. Ibid., 107, 114.

24. Ibid., 115.

25. Albers, ed., *Maria Mitchell,* 106.

26. Maria Mitchell, "Maria Mitchell's Reminiscences of the Herschels," *Century* (1891): 903, 905.

27. Kendall, ed., *Life, Letters,* 108–9.

28. Mitchell, "Reminiscences," 905.

29. Ibid., 908.

30. Kendall, ed., *Life, Letters,* 126.

31. Agnes M. Clerke, *The Herschels and Modern Astronomy* (London: Cassell and Co., 1901), 123–24.

32. Ibid., 124.

33. Ibid., 124–25.

34. Mitchell, "Reminiscences," 909.

35. Ibid., 908.

36. Ibid.

37. Laura Dassow Walls, *Emerson's Life in Science* (Ithaca, NY: Cornell University Press, 2003), 57.

38. Kendall, ed., *Life, Letters,* 128.

Chapter 7: The Yankee Corinnes

1. Phebe Mitchell Kendall, ed., *Maria Mitchell: Life, Letters, and Journals*(1896; repr., Freeport, NY: Books for Libraries Press, 1971), 140.

2. Margaret Fuller, "These Sad but Glorious Days," *Dispatches from Europe,* Larry J. Reynolds and Susan Belasco Smith, eds. (New Haven: Yale University Press, 1991), 108.

3. Kendall, ed., *Life, Letters,* 143.

4. Ibid., 142.

5. Henry Albers, ed., *Maria Mitchell: A Life in Journals and Letters* (Clinton Corners, NY: College Avenue Press, 2001), 113.

6. Harriet Beecher Stowe, *Sunny Memories of Foreign Lands,* vol. 1 (Boston: Phillips, Sampson, 1854), 144.

7. Albers, ed., *Maria Mitchell,* 112.

8. Ibid., 111.

9. Ibid., 113.

10. Stowe, *Sunny Memories,* 147.

11. Ibid., 170.

12. Maria Mitchell Association, Maria Mitchell Papers (MMP), vol. 16; November 24, 1854.

13. George Gordon, Lord Byron, *Childe Harold's Pilgrimage,* available at www.gutenberg.org/etext/5131.

14. Ibid.

15. Richard Brodhead, introduction to *The Marble Faun,* by Nathaniel Hawthorne (New York: Penguin, 1990), xiii.

16. Nathaniel Hawthorne, *Centenary Edition of the Works of Nathaniel Hawthorne,* vol. 14, *The French and Italian Notebooks,* ed. Thomas Woodson (Columbus: Ohio State University Press, 1980), 17.

17. Ibid., 51.

18. Rose Hawthorne Lathrop, *Memories of Hawthorne* (Boston: Houghton Mifflin, 1897), 365.

19. Albers, ed., *Maria Mitchell,* 117–18.

20. Hawthorne, *Notebooks,* 155, 156.

21. Ibid., 156, 157.

22. Kendall, ed., *Life, Letters,* 92.

23. Ibid., 93.

24. Nathaniel Hawthorne, *The Marble Faun,* 35.

25. Maria Mitchell, "Mary Somerville," *Atlantic Monthly* (May 1860): 569.

26. Kendall, ed., *Life, Letters,* 90.

27. Ibid.

28. Hawthorne, *Notebooks,* 157.

29. Hawthorne, *The Marble Faun,* 53.

30. Ibid., 74.

31. T. Walter Herbert, *Dearest Beloved* (Berkeley: University of California Press, 1993), 231.

32. Kendall, ed., *Life, Letters,* 149–50.

33. Hawthorne, *Notebooks,* 158–59.

34. Sophia Hawthorne, *Notes in England and Italy* (New York: G.P. Putnam and Son, 1869), 265.

35. Nathaniel Hawthorne, *Centenary Edition of the Works of Nathaniel Hawthorne,* vol. 18, *The Letters, 1857–1864,* Thomas Woodson and others, eds. (Columbus: Ohio State University Press, 1987), 597.

36. Sophia Hawthorne, *Notes,* 235.

37. Kendall, ed., *Life, Letters,* 92.

38. Hawthorne, *Letters, 1857–1864,* 138–39.

39. Sophia Hawthorne, *Notes,* 272, 274.

40. Kendall, ed., *Life, Letters,* 171.

41. Ibid., 157.

42. Ibid., 151.

43. Ibid.

44. Ibid., 151–52.

45. Albers, ed., *Maria Mitchell,* 121.

46. Ibid.

47. Kendall, ed., *Life, Letters,* 155.

48. Ibid., 156.

49. Albers, ed., *Maria Mitchell,* 122–23.

Chapter 8: A Mentor in Florence

1. Fredrika Bremer, *Bremer's Works,* Mary Howitt, trans. (New York: Harper, 1844), 2: 23.

2. Nathaniel Hawthorne, *Centenary Edition of the Works of Nathaniel Hawthorne,* vol. 14, *The French and Italian Notebooks,* ed. Thomas Woodson (Columbus: Ohio State University Press, 1980), 405–6.

3. Sophia Hawthorne, *Notes in England and Italy* (New York: G.P. Putnam and Son, 1869), 485.

4. Bremer, *Works,* 2:32

5. Hawthorne, *Notebooks,* 435.

6. Maria Mitchell, "The Astronomical Science of Milton: As Shown in 'Paradise Lost,'" *Poet-Lore* 6 (1894): 314.

7. Ibid., 313.

8. Ibid., 320.

9. Ibid., 321–22.

10. Ibid., 322.

11. John Milton, *Paradise Lost,* 1.287–91.

12. Ibid., 5.261–63.

13. Mitchell, "Astronomical Science of Milton," 322.

14. Ibid., 323.

15. Milton, *Paradise Lost,* 5.261–62.

16. Mitchell, "Astronomical Science of Milton," 313.

17. Ibid., 314.

18. Phebe Mitchell Kendall, ed., *Maria Mitchell: Life, Letters, and Journals* (1896; repr., Freeport, NY: Books for Libraries Press, 1971), 151.

19. Elizabeth Barrett Browning, *Aurora Leigh,* 5.183–90.

20. Ibid., 9.941–49.

21. Mary King Babbitt, *Maria Mitchell As Her Students Knew Her* (Poughkeepsie, NY: Enterprise Publishing, 1912), 19–20.

22. Maria Mitchell, "Mary Somerville," *Atlantic Monthly* (May 1860): 568.

23. Ibid., 570.

24. William Whewell, "Review of 'On the Connexion of the Physical Sciences,'" *Quarterly Review* 51 (1834): 68.

25. Ibid., 59.

26. Ibid., 65–66.

27. Ibid., 65.

28. Ibid., 56.

29. John Dryden, *Selected Poems,* Steven N. Zwicker, ed. (New York: Penguin, 2002), 295.

30. Whewell, "Review," 68.

31. Kathryn A. Neeley, *Mary Somerville: Science, Illumination and the Female Mind* (Cambridge, UK: Cambridge University Press, 2001), 9.

32. Whewell, "Review," 68.

33. Neeley, *Mary Somerville,* 78.

34. Ibid.

35. Ibid.

36. Ibid., 11.

37. Mitchell, "Mary Somerville," *Atlantic Monthly* (May 1860), 571.

38. Kendall, ed., *Life, Letters,* 160.

39. Julia Ward Howe, "Maria Mitchell," *Our Famous Women* (Hartford, CT: A. D. Worthington, 1884), 452.

40. Kendall, ed., *Life, Letters,* 162.

41. Henry Albers, ed., *Maria Mitchell: A Life in Journals and Letters* (Clinton Corners, NY: College Avenue Press, 2001), 124.

42. Ibid., 125.

43. Ibid.

44. Mitchell, "Mary Somerville," 568.

Chapter 9: The War Years

1. C. F. Brooks, "Island Nantucket," *Geographical Review* 4, no. 3 (September 1917): 197–207 (population chart on p. 202).

2. Phebe Mitchell Kendall, ed., *Maria Mitchell: Life, Letters, and Journals* (1896; repr., Freeport, NY: Books for Libraries Press, 1971), 167.

3. Helen Wright, *Sweeper in the Sky: The Life of Maria Mitchell* (1949; repr., Clinton Corners, NY: College Avenue Press, 1997), 71.

4. Ibid., 128.

5. "Maria Mitchell," *Emerson's United States Magazine* 5, no. 37 (July 1857): 94.

6. "Miss Mitchell's Telescope and Observatory," *Scientific American* 1, no. 15 (October 8, 1859): 234.

7. Margaret Moore Booker, *The Admiral's Academy* (Nantucket: Mill Hill Press, 1998), 34–36.

8. Henry Albers, ed., *Maria Mitchell: A Life in Journals and Letters* (Clinton Corners, NY: College Avenue Press, 2001), 134–35.

9. Ibid.

10. www.nha.org/hn/HNracerelations.html.

11. Wright, *Sweeper,* 97.

12. *Nantucket Inquirer,* January 16, 1841.

13. *Nantucket Inquirer,* July 31, 1841.

14. Augusta Jane Evans, *Macaria; or, Altars of Sacrifice,* ed. Drew Gilpin Faust (Baton Rouge: Louisiana State University Press, 1992), 174.

15. Evans, *Macaria,* 179–80.

Chapter 10: Vassar Female College

1. Elizabeth Daniels, *Main to Mudd, and More* (Poughkeepsie, NY: Vassar College, 1996), 12.

2. Helen Lefkowitz Horowitz, *Alma Mater* (Boston: Beacon Press, 1984), 30.

3. *Godey's Lady's Book* (January 1864), 93.

4. Matthew Vassar, *The Autobiography and Letters of Matthew Vassar,* ed. Elizabeth Hazelton Haight (New York: Oxford University Press, 1916), 42.

5. *Godey's Lady's Book* (January 1864): 93.

6. Barbara Miller Solomon, *In the Company of Educated Women* (New Haven, CT: Yale University Press, 1985), 24.

7. Lynn Peril, *College Girls: Bluestockings, Sex Kittens, and Co-Eds, Then and Now* (New York: W. W. Norton, 2006), 43.

8. Stanley M. Guralnick, *Science and the Ante-Bellum American College* (Philadelphia: American Philosophical Society, 1975), 7.

9. Ibid., 13.

10. Ibid., 15.

11. Kim Tolley, *The Science Education of American Girls* (New York: RoutledgeFalmer, 2003), 35.

12. Guralnick, *Science and the Ante-Bellum American College,* 30–32.

13. Peril, *College Girls,* 35.

14. Miriam R. Levin, *Defining Women's Scientific Enterprise* (Lebanon, NH: University Press of New England, 2005), 3.

15. Ibid., 23.

16. Daniels, *Main to Mudd,* 15.

17. Peril, *College Girls,* 39.

18. Sarah Josepha Hale, *Godey's Lady's Book* (May 1864): 489.

19. Karen Sànchez-Eppler, "Bodily Bonds: The Intersecting Rhetoric of Feminism and Abolition," in Shirley Samuels, ed., *The Culture of Sentiment* (New York: Oxford University Press, 1992), 92.

20. Solomon, *In the Company,* 50.

21. Peril, *College Girls,* 41.

22. Olivia Mancini, "Vassar's First Black Graduate: She Passed for White," *Journal of Blacks in Higher Education,* no. 34 (Winter 2001–2002), 108–9. Interestingly, Vassar and Spelman were both supported by generous donations from John D. Rockefeller. Rockefeller was Spelman's primary benefactor, and the name of the school was changed from Atlanta to Spelman to honor Rockefeller's wife. It is obvious that members of the Vassar community knew that black women were perfectly capable of college-level work and that some supported them in their efforts—but there were few challenges to the segregationist model. One black student, Anita Florence Hemmings, did attend Vassar in the nineteenth century, but she did so by successfully passing for white; her racial identity was exposed shortly before her graduation in 1897, at which point controversy broke out.

23. Londa Schiebinger, *The Mind Has No Sex? Women in the Origins of Modern Science* (Cambridge, MA: Harvard University Press, 1989), 245.

24. Sally Gregory Kohlstedt, "Maria Mitchell: The Advancement of Women in Science," *New England Quarterly* 51, no. 1 (March 1978): 39.

25. Cynthia Eagle Russett, *Sexual Science: The Victorian Construction of Womanhood* (Cambridge, MA: Harvard University Press, 1989), 3–4.

26. Nancy Leys Stepan, "Race and Gender: The Role of Analogy in Science," *Isis* 77, no. 2 (June 1986): 268.

27. Louise M. Newman, *White Women's Rights: The Racial Origins of Feminism in the United States* (New York: Oxford University Press, 1999), 4–5.

28. Henry Albers, ed., *Maria Mitchell: A Life in Journals and Letters* (Clinton Corners, NY: College Avenue Press, 2001), 161–62.

Chapter 11. No Miserable Bluestocking

1. Henry Albers, ed., *Maria Mitchell: A Life in Journals and Letters* (Clinton Corners NY: College Avenue Press, 2001), 141–42.

2. Maria Mitchell, "Mary Somerville," *Atlantic Monthly* (May 1860), 569.

3. Lynn Peril, *College Girls: Bluestockings, Sex Kittens, and Co-Eds, Then and Now* (New York: W. W. Norton, 2006), 28–33.

4. Ibid., 29.

5. Albers, ed., *Maria Mitchell,* 144–45.

6. Horatio Hale, *Godey's Lady's Book* (February 1864): 199.

7. Sarah Josepha Hale in ibid., May 1865, 464.

8. Madeleine Stern, *We the Women* (Lincoln: University of Nebraska Press, 1994), 147.

9. *Godey's Lady's Book,* June 1864.

10. Matthew Vassar, *The Autobiography and Letters of Matthew Vassar,* ed. Elizabeth Hazelton Haight (New York: Oxford University Press, 1916), 135–36.

11. Albers, ed., *Maria Mitchell,* 150–51.

12. Ibid., 153.

13. Ibid., 200.

14. *Boston Globe,* January 17, 1873.

15. Linda Babcock and Sara Laschever, *Women Don't Ask* (Princeton, NJ: Princeton University Press, 2003), 6–7.

16. Chris M Golde, "After the Offer, Before the Deal: Negotiating a First Academic Job," *Academe* 85, no. 1 (January 1, 1999): 46.

17. Vassar, *Autobiography,* 137.

18. Helen Lefkowitz Horowitz, *Alma Mater* (Boston: Beacon Press, 1984), 32.

19. Albers, ed., *Maria Mitchell,* 165.

20. Elizabeth Daniels, *Main to Mudd, and More* (Poughkeepsie, NY: Vassar College, 1996), 16.

21. Albers, ed., *Maria Mitchell,* 163.

22. Maria Mitchell Papers (MMP), Archives and Special Collections, Vassar College Libraries.

23. Maria Mitchell Association, MMP, vol. 16; 1853.

Chapter 12: "Good Woman That She Is"

1. Henry Albers, ed., *Maria Mitchell: A Life in Journals and Letters* (Clinton Corners, NY: College Avenue Press, 2001), 158.

2. Mary King Babbitt, *Maria Mitchell As Her Students Knew Her* (Poughkeepsie, NY: Enterprise Publishing, 1912), 19–20; Helen Wright, *Sweeper in the Sky: The Life of Maria Mitchell* (1949; repr., Clinton Corners, NY: College Avenue Press, 1997), 169.

3. Ibid.

4. Babbitt, *Students,* 25.

5. Maria Mitchell Papers, Archives and Special Collections, Vassar College Libraries (VCL).

6. Babbitt, *Students,* 26.

7. VCL.

8. Ibid.

9. Albers, ed., *Maria Mitchell,* 164.

10. VCL.

11. Phebe Mitchell Kendall, ed., *Maria Mitchell: Life, Letters, and Journals* (1896; repr., Freeport, NY: Books for Libraries Press, 1971), 174–75.

12. Miriam Levin, *Defining Women's Scientific Enterprise: Mount Holyoke Faculty and the Rise of American Science* (Lebanon, NH: University Press of New England, 2005), 26–27.

13. Albers, ed., *Maria Mitchell,* 170.

14. Kendall, ed., *Life, Letters,* 180.

15. Ibid.

16. Ibid., 178.

17. Albers, ed., *Maria Mitchell,* 171.

18. John Lankford, *American Astronomy: Community, Careers, and Power, 1859–1940* (Chicago: University of Chicago Press, 1997), 300.

19. Ibid., 299.

20. Albers, ed., *Maria Mitchell,* 160–61.

21. Ibid., 161.

22. Kendall, ed., *Life, Letters,* 187.

23. Ibid., 184.

24. Albers, ed., *Maria Mitchell,* 198–99.

25. Ibid., 221.

26. Ibid., 239.

27. Ibid., 195.

28. "The Female Astronomer: History of Miss Maria Mitchell of Vassar," *New York Times,* August 9, 1881.

29. Anne Mitchell, "Reminiscences of the Family Life," *Friend's Intelligencer,* December 21, 1889.

30. Wright, *Sweeper in the Sky,* 173.

Chapter 13: The Undevout Astronomer

1. Henry Albers, ed., *Maria Mitchell: A Life in Journals and Letters* (Clinton Corners, NY: College Avenue Press, 2001), 159.

2. Ibid.

3. Ibid., 254.

4. Ibid., 186.

5. Louisa May Alcott, *Work: A Story of Experience* (1813; repr., New York: Penguin, 1994), 157.

6. Albers, ed., *Maria Mitchell,* 287.

7. Anne Mitchell Macy, "Astronomical Garret" (Maria Mitchell Association), n.p., n.d.

8. Helen Wright, *Sweeper in the Sky: The Life of Maria Mitchell* (1949; repr., Clinton Corners, NY: College Avenue Press, 1997), 11.

9. Alcott, *Work,* 213.

10. Ibid., 442.

11. John Frederic Meyer, *History of the Unitarian Church of Nantucket* (Nantucket, MA: Inquirer and Mirror, 1902), 9.

12. Henry F. Edes, *An Abstract of Unitarian Belief* (Boston: printed but not published by Freeman and Bolles, 1836), 52.

13. Ralph Waldo Emerson, "Historic Notes of Life and Letters in New England," *Atlantic Monthly* (1883); reprinted in *The Portable Emerson,* ed. Carl Bode (New York: Penguin, 1981), 600; Laura Dassow Walls, *Emerson's Life in Science* (Ithaca, NY: Cornell University Press, 2003), 83.

14. Albers, ed., *Maria Mitchell,* 222.

15. Phebe Mitchell Kendall, ed., *Maria Mitchell: Life, Letters, and Journals* (1896; repr., Freeport, NY: Books for Libraries Press, 1971), 188.

16. Albers, ed., *Maria Mitchell,* 46.

17. Ibid., 287.

Chapter 14: Retrograde Motion

1. Edward H. Clarke, *Sex in Education, or, A Fair Chance for Girls* (Boston: James R. Osgood, 1874), 139.

2. Ibid., 45, 93.

3. Ibid., 14.

4. Ibid., 115.

5. Cynthia Eagle Russett, *Sexual Science: The Victorian Construction of Womanhood* (Cambridge, MA: Harvard University Press, 1989), 12.

6. Ibid., 206.

7. Foucault, *The History of Sexuality: An Introduction* (New York: Vintage, 1980).

8. Henry Albers, ed., *Maria Mitchell: A Life in Journals and Letters* (Clinton Corners, NY: College Avenue Press, 2001), 232.

9. Ibid., 231.

10. Julia Ward Howe, *Sex and Education: A Reply to Dr. E. H. Clarke's "Sex in Education"* (Boston: Roberts Bros., 1874), 24.

11. Albers, ed., *Maria Mitchell,* 233.

12. Julia Ward Howe, *Reminiscences, 1819–1899* (Boston: Houghton Mifflin, 1899), 385, 387.

13. Maria Mitchell Association (MMA) box.

14. Maria Mitchell, "The Higher Education of Women," Congress of Women 1874, http://nrs.harvard.edu/um-3:FHCL:507525.

15. Phebe Mitchell Kendall, ed., *Maria Mitchell: Life, Letters, and Journals* (1896; repr., Freeport, NY: Books for Libraries Press, 1971), 215.

16. Maria Mitchell, "Address of the President," Congress of Women 1874, http://nrs.harvard.edu/um-3:FHCL:507525.

17. Ibid., 4.

18. Kendall, ed., *Life, Letters,* 187.

19. Maria Mitchell, "Report of the Committee on Science," Congress of Women 1874, http://nrs.harvard.edu/um-3:FHCL:507529.

20. Albers, ed., *Maria Mitchell,* 238.

21. Maria Mitchell Papers, Archives and Special Collections, Vassar College Libraries (VCL).

22. Ibid.

23. John Lankford, *American Astronomy: Community, Careers, and Power, 1859–1940* (Chicago: University of Chicago Press, 1997), 325.

24. Pamela Mack, "Women in Astronomy in the United States, 1875–1920," senior thesis, Harvard University, 1977.

25. VCL.

26. Dorrit Hoffleit, *Women in the History of Variable Star Astronomy* (Cambridge, MA: American Association of Variable Star Observers, 1993), 3.

27. Mack, "Women in Astronomy," 79.

28. Ibid., 80.

29. VCL.

Chapter 15: Urania's Inversion

1. Henry Albers, ed., *Maria Mitchell: A Life in Journals and Letters* (Clinton Corners, NY: College Avenue Press, 2001), 172.

2. Ibid., 187.

3. Phebe Mitchell Kendall, ed., *Maria Mitchell: Life, Letters, and Journals* (1896; repr. Freeport, NY: Books for Libraries Press, 1971), 252.

4. Ibid., 187–88.

5. John Lankford, *American Astronomy: Community, Careers, and Power, 1859–1940* (Chicago: University of Chicago Press, 1997), 305.

6. Ella Rodman Church, "Maria Mitchell," *Godey's Lady's Book,* November 1879.

7. Albers, ed., *Maria Mitchell,* 283.

8. Ibid.

9. Ibid., 302.

10. Ibid., 312.

11. Ibid., 314.

12. Ibid., 320.

13. Ibid., 323.

14. Graceanna Lewis (1889), "A Reminiscence of Maria Mitchell," quoted in *Historic Nantucket,* January 1987 34:3, 24.

15. Martin, "Melville and Sexuality," in *The Cambridge Companion to Herman Melville,* ed. Robert S. Levine (New York: Cambridge University Press, 1998), 186.

16. Leon Howard, *Herman Melville, a Biography* (Berkeley: University of California Press, 1951), 334.

17. Nathaniel Philbrick, "Hawthorne, Maria Mitchell, and Melville's 'After the Pleasure Party,'" *ESQ: A Journal of the American Renaissance,* 37, no. 4 (1991): 294.

18. Herman Melville, *The Poems of Herman Melville,* ed. Douglas Robillard (Kent, OH: Kent State University Press, 2000), 311.

19. Ibid., 311.

20. Ibid., 314.

21. Gillian Beers, "'Authentic Tidings of Invisible Things': Vision and the Invisible in the Later Nineteenth Century," in *Vision in Context: Historical and Contemporary Perspectives on Sight,* eds. Teresa Brennan and Martin Jay (New York: Routledge, 1996), 88.

22. Albers, ed., *Maria Mitchell,* 219.

23. Lankford, *American Astronomy,* 305.

24. Maria Mitchell Papers, Archives and Special Collections, Vassar College Libraries.

25. Lewis, *Historic Nantucket,* 24.

26. Margaret Fuller, *Woman in the Nineteenth Century* (London: H. G. Clarke, 1845), 109, 108.

27. Kendall, ed., *Maria Mitchell,* 187.

Epilogue

1. In his "Remarks at National Bureau of Economic Research Conference on Diversifying the Science and Engineering Workforce," Summers said, "... It does appear that on many, many different human attributes—height, weight, propensity for criminality, overall IQ, mathematical ability, scientific ability—there is relatively clear evidence that whatever the difference in means—which can be debated—there is a difference in the standard deviation, and variability of a male and a female population. And that is true with respect to attributes that are and are not plausibly, culturally determined..." For Summers's full speech and later accounts, see www.president.harvard.edu/speeches/2005/nber.html and www.president.harvard.edu/speeches/2005/womensci.html. For a taste of the resulting controversy, see the *Boston Globe* starting January 17, 2005.

2. Donna E. Shalala, "Dedication," from *Beyond Bias and Barriers: Fulfilling the Potential of Women in Academic Science and Engineering* (Washington, DC: National Academies Press, 2007), v.

3. Paul Fain, "Too Much, Too Fast," *Chronicle of Higher Education,* January 19, 2007.

4. Summers, "Remarks at NBER Conference," www.president.harvard.edu/speeches/2005/nber.html.

5. *Beyond Bias and Barriers,* 41.

6. Ibid., 42.

7. Ibid., 2.

8. Ibid., 1.

9. Maria Mitchell Association (MMA), box 17.

10. Anna C. Brackett, "Maria Mitchell," *Century Illustrated Magazine* 37, no. 6 (October 1889): 954.

11. "The Female Astronomer: History of Miss Maria Mitchell of Vassar," *New York Times,* August 9, 1881.

12. *Beyond Bias and Barriers,* 4.

13. Emily Davies, *The Higher Education of Women* (London: Alexander Strahan, 1866), 133.

14. Ella Rodman Church, "Maria Mitchell," *Godey's Lady's Book,* November 1879.

15. Maria Mitchell Papers, Archives and Special Collections, Vassar College Libraries (VCL).

16. Thomas Wentworth Higginson, "Maria Mitchell," in *Woman,* ed. William C. King (Springfield, MA: King-Richardson Company, 1902), 391.

17. Marcus Benjamin, "Obituary of Maria Mitchell," *Scientific American* (July 20, 1889): 38.

18. Harriet Stanton Blatch, "Voluntary Motherhood," in *Transactions of the National Council of Women of the United States,* Rachel Avery, ed. (New York: Lippincott, 1891), 283.

19. Edouard A. Stackpole, foreword, in *Two Steps Down,* Alice Albertson Shurrocks (Nantucket, MA: Nantucket Inquirer and Mirror Press, 1953), n.p.

20. C. P. Snow, *The Two Cultures and the Scientific Revolution* (Cambridge 1993), 103. Cited by Jan Golinski, "The Care of the Self and the Masculine Birth of Science," *History of Science* 40, pt. 2, no. 128 (June 2002): 125.

21. Betty Friedan, *The Feminine Mystique* (New York: W. W. Norton, 1963, 2001), 153.

22. Albers, ed., *Maria Mitchell,* 327.

MM in index refers to Maria Mitchell. WM in index refers to William Mitchell.

AAW. *See* Association for the Advancement of Women (AAW)

abolitionist movement, 156, 159–63, 197, 220. *See also* slavery

Adams, John Couch, 101, 106, 108, 114, 116

Adams, John Quincy, 65–66

African Americans: analogous relationship between women as inferior sex and, 174–75; education of, 172; employment of, 156; equality for, 232–34; segregation of, 233; vote for, 156, 233. *See also* slavery

"After the Pleasure Party" (Melville), 243–47

Agassiz, Louis, 70

Agnesi, Maria, 148

Airy, George, 34, 92, 97–98, 101–5, 111, 114

Airy, Richarda, 102, 105, 110, 150

Akers, Paul, 154

Albers, Henry, 33–34, 189

Alcott, Bronson, 202, 203, 214, 215–16

Alcott, Louisa May, 195, 213, 215–16, 221

American Academy of Arts and Sciences, xiv, 68, 138, 231

American Association for the Advancement of Science, 70

American Association of University Professors, 186

American Astronomical Society, 63

American Journal of Science and Arts, 56

American Philosophical Society, 70, 209, 231

Amherst College, 28, 172

Amici, Giovanni Battista, 140

Antioch College, 168, 182

Aristotle, 132

Arnold, Matthew, 241

Arnott, Neil, 104

Association for the Advancement of Women (AAW), xiv, 227, 228, 229, 230, 231, 256

astronomical clock, 15, 16, 22–24

Astronomische Nachrichten, 56

astronomy: American opportunities for,

in nineteenth century, 108–9; collaborative nature of, 43–44, 47; comets' significance for, 57, 62–64; and eclipse observations, 18–19, 208–10; and Emerson, 10, 33, 220; and mathematics, xi–xii, 43, 69–70; in Melville's "After the Pleasure Party," 245–46; and Milton, 59, 84, 140, 142; Nantucketers' interest in, 7–10, 14; and planetary motion, xi–xiv, xvii; and poetry, xvii, 38–39, 142, 190–91; salaries for astronomers, 64, 65; WM as astronomer, 3–4, 7–12, 18–19, 24, 42; women astronomers, xvi, 3–4, 5, 42–43, 44, 47–48, 59–61, 83, 110–14; women professors in, 183; women's study of, as important, 8–9. *See also* comets; England; Europe; France; Herschel, Caroline; Italy; Mitchell, Maria; observatories; telescopes; *and specific astronomers*
athenaeum movement, 21, 29–34. *See also* Nantucket Atheneum
Atlanta College, 172
Atlantic Monthly, 85, 125, 149–50, 152, 155, 243
Aurora Leigh (Barrett Browning), 91, 92, 123, 143–44
Austria, 93, 119
Avery, Alida, 227

Babbage, Charles, 104–5, 114
Babbitt, Mary King, 144, 195, 196, 248, 249
Babcock, Linda, 186
Babcock, Rufus, 165, 177–80, 183–84
Bache, Alexander Dalles, 27, 56, 66, 68
Baden-Powell, Mr., 104
Ballou, Adin, 30–31
Bancroft, George, 66
Baptists, 218
Barnard, Lilla, ix–xi, xii–xiii, xvii, xviii

Barrett Browning, Elizabeth: compared with Milton, 143–44; and Fuller, 123; MM's failure to meet with, 144–45; MM's stories on, to Vassar students, 205; Poe's review of poetry by, 86; poetry by, 86, 91, 92, 105, 112, 123, 143–44; as role model for MM, 156; travels by, in Europe, 116, 117, 119–21; Whewell on poetry by, 105
Bates College, 172
Baudelaire, Charles, 117
Baym, Nina, 38
Beatrice Cenci (Hosmer), 127
Beers, Gillian, 248
Benton, Thomas Hart, 127
Beyond Bias and Barriers, 250–54
Bishop, Nathan, 212
blacks. *See* African Americans
Blatch, Harriet Stanton, 256
The Blithedale Romance (Hawthorne), 121, 122
bluestockings, 178–79
Bond, George: as assistant to William Bond, 27, 65; astronomical observations by, 54; discovery of comets by, 67; at Harvard Observatory, 27, 32, 65, 67, 103, 200; and MM's discovery of comet, 55–56, 67; and MM's trip to Europe, 92; as professor of astronomy at Harvard, 65; telescope built by, 47; on Troy Female Seminary, 165
Bond, William: discovery of comet by, 54; at Harvard Observatory, 27, 32, 64–65, 67, 103; and MM's discovery of comet, 55, 67, 68; and MM's trip to Europe, 92; telescope built by, 47
Bonheur, Rosa, 116, 117, 205
Booth, Lydia, 167
Boston, 98
Boston Athenaeum, 29–30
Boston Globe, 185

Bowdoin College, 172
Brackett, Anna, 253
Bremer, Fredrika, 137, 139, 140
Britain. *See* England; Scotland
British Association for the Advance-
ment of Science, 105
Brodhead, Richard, 121
Brontë, Charlotte, 91
Brook Farm, 85–86
Brown, Helen, 248–49
Brown, John, 159, 160, 197
Browning, Robert, 121
Brownson, Orestes, Jr., 85–86
Brown University, 85
Buell, Lawrence, 245
Burney, Fanny, 59–60, 112
Byron, George Gordon, Lord, 91, 98,
119–21, 130

Caesar, Julius, 58–59, 121
Cahan, David, 49
Cambridge observatory, 107
Cambridge University, xv, 67, 101,
105–7, 145
Catholic Church, 132–34, 141, 143
Cavendish, Margaret, 46
Cenci, Beatrice, 124, 126, 127
Century, 112
Challis, Professor, 107
Channing, William, 219
Charles, King, 59
chemistry, 237, 238
Cheney, Ednah, 221
Child, Lydia Maria, 30, 163, 232–33
Childe Harold (Byron), 91, 120
Christian Examiner, 225–26
Christina, Queen, 46
Chronicle of Higher Education, 251–52
chronometers, 5, 12, 19, 34, 97
Cicero, 121
Civil War, 94, 156, 163–65, 171, 173,
210, 216–17, 224
Clairault, 62

Clark, Alvan, 158
Clarke, Edward, xv, 224–30
Clarke, James Freeman, 75–76, 118,
120, 130
Clerke, Agnes Mary, 111–12
Coastal Survey, U.S., 11–12, 27, 56, 64,
66, 70, 208
Coffin, Charles C., 162, 209
Coffin, Mary S., 17–18
Coffin, Rebecca, 254
Coleman, Lydia. *See* Mitchell, Lydia
Coleman
Coleman, Lydia Wing, 16–17, 18
Coleman, Phebe Folger, 1, 15–16,
18, 19
Coleridge, Samuel Taylor, 146
Collège de France, 47
colleges and universities: curricula of,
28, 168–70; dangers of college-level
education for women, 224–28, 233;
men's colleges, 169–70; MM's and
Howe's defense of higher education
for women, 226–30; in 1900, 52,
135–36; opposition to women's
study of science at, 234–38; publications by faculty at, 206; salaries in,
184–87, 251; science curriculum
in, 175–76; women professors at,
177–78, 180–87, 253; women's
colleges, 135–36, 166–68, 170,
175–76. *See also* education; Harvard
College/University; Vassar College;
women's education; *and other specific
colleges and universities*
Columbia University, 238
comets: William Bond's discovery of,
54; George Bond's discovery of, 67;
Donati's discovery of, 140; great
comet (1843), 65; Hale-Bopp comet,
61; Halley's comet, 48, 58, 60, 62;
Caroline Herschel's discovery of,
3–4, 43, 59–60, 63, 112; MM's discovery of, xvi, 53–58, 63–64, 67,

70–71; mysterious nature of, 61; orbits of, 48, 56, 57, 61, 62; Shoemaker-Levy 9 comet, 61; significance of, for astronomy, 57, 62–64; symbolism of, 57–60; telescope for discovery of, 60–61, 64; women's discoveries of, 59–60. *See also* astronomy

Congress of Women, 227, 228–29

Conway, Ann, 46

Cooper, Arthur, 161, 162

Copernicus, 220, 240

Cornell University, 204

Coterie literary club, 35–37

Crèvecoeur, Hector St. John de, 79–80

The Cricket on the Hearth (Dickens), 91

Cromwell, Oliver, 59

Cunitz, Marie, 47

Curie, Marie, 256, 257

Dame, Kate Mitchell, 2, 153, 241, 242

Daniels, Elizabeth, 170

Dante, 141

Dartmouth College, 204

Darwin, Charles, xiv–xv, 102, 114, 144, 171, 172–73, 174

Davis, Charles Henry, xi–xii, 69

Dean, Rebecca Pennell, 182

Denton, Denice Dee, 250–52, 258

Descartes, René, 45–46, 49

The Descent of Man, and Selection in Relation to Sex (Darwin), 173

De Staël, Madame, 115, 118–21

Dial, 76

Dickens, Charles, 91, 92, 98, 121

Dickinson, Anna, 221

Dickinson, Emily, 38–39, 40, 213

"Discourse on Woman" (Mott), xv–xvi

Dix, Dorothea, 83–84, 85, 240

domestic ideology, xii–xiii, 125, 156, 178, 179–80, 187

Donati, Giovanni Battista, 140

Douglass, Frederick, 10, 30, 33, 162–63, 232–33

Dryden, John, 147–48

Earle, Lizzie, 83, 88

eclipse observations, 18–19, 208–10. *See also* astronomy

Edes, Henry, 220

Edgeworth, Maria, 1–2, 5, 21, 51–52, 149

Edgeworth, William, 51

Edinburgh Review, 178–79

education: of African Americans, 172; and athenaeum movement, 21, 29–34; in eighteenth century, 22, 168–69; free public education in nineteenth century, 21, 25–26; home education of children, 4–5, 9–10, 22; of males, 13, 27, 28, 168–70; Massachusetts law on public education, 25; Nantucket schools, 4, 10, 15, 25–29; and Quakers, 25–26, 256; and salaries for teachers, 25, 26; scientific education in the home, 4–5, 9–10, 22; secondary education, 27, 52, 169, 255–56; and women teachers, 25–27. *See also* colleges and universities; women's education; *and specific colleges and universities*

Edwards, Jonathan, 28, 48

Edward the Confessor, 58

Einstein, Alfred, 99

Eliot, George, 91

Elizabeth, Queen, 100

Elizabeth of Bohemia, Princess, 46

Elmira Female College, 183

Emerson, Ralph Waldo: and astronomy, 10, 33, 220; death of, 241; on Margaret Fuller, 118; and John Herschel's *Preliminary Discourse,* 113; as influence on MM, 203; recommendation from, for Vassar student, 195; reference to, in MM's poem for her student, 195; on science, 40; Toynbee on, 107; and Unitarianism, 219; Whewell on, 105

Emerson's United States Magazine, x, 157–58

Encke, Johann Franz, 93

England: astronomers and other scientists in, 92, 96–97, 99–108, 110–14; Cambridge University in, xv, 67, 101, 105–7; Greenwich observatory in, 66–67, 97–98, 102–4; Lake District of, 105; Liverpool in, 94–98; London in, 98–102, 104–5; Lunar Society in, 101; Manchester in, 96; MM's travels in, 91–115; MM's trip to, 94; national mind-set of, 107–9; observatories in, 64, 66–67, 96–98, 102–4, 107; women's education in, 21, 115, 149. *See also* Royal Society

Europe: expense and difficulty of travel in, 120–21; Margaret Fuller on Americans' need to travel to, 92; MM in England, 91–93, 94–115; MM in France, 115–19; MM in Italy, 121–52; MM's traveling companion to, 94, 109–10; MM's trip to, 94; observatories in generally, 64, 116; reasons for MM's trip to, 92–94; science and astronomy in, 47–48, 56–57, 64, 92–93, 96–97, 99–108, 108, 110–14, 116, 133–35, 140, 144–52, 174. *See also* England; France; Italy; *specific countries*

Evans, Augusta, 164–65

Everett, Edward, 55, 56–57, 67–68, 92

Faraday, Michael, xiv–xv, 101, 105, 147

Farrar, Charles, 183, 186–87, 189

Faust, Drew Gilpin, 258

female education. *See* women's education

femaleness. *See* sexual identity; women

feminism. *See* women's rights

Fisk College, 172

Flamsteed, John and Margaret, 47

Les Fleurs du Mal/The Flowers of Evil (Baudelaire), 117

Florence, Italy, 137–52

Folger, Abiah, 14

Folger, Anna, 15

Folger, Elisabeth Starbuck, 15

Folger, Phebe. *See* Coleman, Phebe Folger

Folger, Walter: and astronomical clock, 15, 16, 22–24; college not attended by, 28; home of, 22; library of, 1, 11, 22; and Nantucket Philosophical Institution, 15, 19, 20, 24, 25, 30; publications by, 22; relationship between WM and, 24–25; scientific achievements of, 15, 16, 22–24; studies of, with sister Phebe, 15, 24; telescope of, 22

Foucault, Michel, 225

Framingham State College, 27

France: astronomers and other scientists in, 47–48, 92, 108, 116; Institut de France in, 19, 64, 117; MM's travels in, 115–19; national mind-set of, 108; observatory in, 116; Paris in, 116–19; sexual freedom in Paris, 117–18

Frankenstein (Shelley), 51, 60

Franklin, Benjamin, x, 14–15, 21, 28, 48, 49

Friedan, Betty, 233, 257

Fuller, Margaret: on Americans' need to travel to Europe, 92; and Barrett Browning, 123; biographers of, 75–76, 227; and Boston seminars for women, 76, 85; death of, 76; and Hawthornes, 121, 122–24, 129; in Italy, 76, 116, 118–21, 130; on marriage, 77; marriage of, 76, 123; in Paris, 116, 117; portrayal of, in Bar-

rett Browning's *Aurora Leigh,* 123;
portrayal of, in Hawthorne's fiction,
121, 122, 123–25; as role model for
MM, 77–78, 84; on women's role
and sexual identity, 75–77, 82, 244;
writings by, 76, 77, 82
Furness, Caroline, 234–35, 237

Galileo: as astronomer, 43, 62–63,
142–43, 220; Catholic Church's
prosecution of, 132–33, 134, 137,
139, 143; grave of, 138; Hawthornes
on, 138–40; home of, near Florence,
138, 140; and Milton, 59, 137, 142,
144; MM on, 127, 142–43; and tele-
scope, 62–63, 142; Tribune memo-
rial to, in Florence, 138–40
Gardner, Anna, 162, 163
Gardner, Oliver, 161, 162
Garrison, William Lloyd, 31, 162–63
Gaston, Mary, 234
gender. *See* sexual identity; women
George, King, 185–86
Germany, 64, 93
Gibson, John, 126
Goddard, Martha, 234
Godey's Lady's Book, 83, 90, 167, 171,
173, 181–82, 201, 254–55
Godwin, William, 50–51
Goethe, Johann Wolfgang von, 121
Golinski, Jan, 45
Grant, Robert, 62
Great Awakening, Second, 218–19
Great Britain. *See* England; Scotland
Greenwich, England, observatory,
66–67, 97–98, 102–4
Grey, Asa, 70
Grimké, Angelina, 82
Guralnick, Stanley, 168–69
Guthrie, James, 39

Hale, Edward Everett, 225–26, 227
Hale, Horatio, 181–82

Hale, Sarah Josepha, 167–68, 171–72,
181–83
Hale-Bopp comet, 61
Halley, Sir Edmund, 62
Halley's comet, 48, 58, 60, 62
Hannaford, Phebe, 83
Harding, Sandra, xiv
Hartnup, John, 96, 97
Harvard College/University: astronomy and mathematics courses at,
176, 195–96, 204; astronomy
research at generally, 93, 200; curriculum of, 28, 168–69, 183; Everett
as president of, 55, 56–57, 67–68,
92; Faust as president of, 258; observatory at, 11, 27, 32, 44, 64–65, 66,
67–68, 103, 193, 236–37; pedagogy
at, 202; professors at, 252; Summers
as president of, 250, 252, 258; Unitarian covenant of, 220; WM's decision not to attend, 2; women as
special students at, 235, 237; women
astronomy assistants at, 44, 236–37;
women excluded from, in nineteenth century, 27
Harvard Medical School, xv, 224
Hawthorne, Nathaniel: children of,
122, 125, 131; and Margaret Fuller,
121, 122–25, 129; on Galileo,
138–40; and Hosmer, 129, 135; Lander's sculpture of, 127; and Melville,
245; and nudity in paintings and
sculpture, 126, 127; personality of,
123, 125, 131–32; physical appearance of, 98; reference to, in MM's
poem for her student, 195; relationship between Sophia Peabody and,
85; in Rome, 121–32, 135–36, 245;
writings by, 121, 122, 123–25, 126,
130, 131–32, 136, 245
Hawthorne, Rose, 122
Hawthorne, Sophia Peabody: children
of, 122, 125, 131; and Fuller, 121,

122–24, 129; on Galileo, 139, 140;
in Italy with MM, 121–27, 129–30,
132, 154; relationship between
Hawthorne and, 85; and Ida
Russell, 85
Hecker, Isaac, 86
Hemans, Felicia, 105
Henry, Joseph, 70, 92
Herbert, T. Walter, 127
Herschel, Caroline: as assistant to
brother William, 3, 5, 42–43, 44, 47,
111–13; clothing of, 110; death of,
110, 138; discovery of comets by,
3–4, 43, 59–60, 63, 112; handwritten
notes of, 111, 113; Lalande's daughter
named for, 48; mathematical
achievements of, 42–43; MM on,
112–13, 185–86; Lucretia Mott on,
xvi; Adrienne Rich on, 248; as role
model for MM, 110; and Royal
Society, 110, 111, 138; salary of,
185–86; work routine of, 111–12
Herschel, John: as astronomer, 34, 61,
101, 113; on comets, 61; daughter
of, 133; MM's meeting with, 92,
110–11, 113, 114, 150; and Royal
Society, 149; and Mary Somerville,
147, 148, 150; writings by, 113
Herschel, Lady Margaret, 150
Herschel, William: astronomy career
of, 42; death of, 110; discovery of
Uranus by, 42; lack of mathematical
skills of, 42; as music conductor, 42;
sister Caroline as assistant to, 3, 5,
42–43, 44, 47, 111–13; on telescopes,
63; writings by, 42
Hevelius, Johannes and Elizabetha,
43, 47
Hewson, William, 15
Higginson, Thomas Wentworth, 256
higher education. *See* colleges and universities; education; *and specific
universities*

Hoffleit, Dorrit, 236–37
Hoffmann, August Wilhelm Von, 101,
102, 114
home economics, 237, 238
Homer, 148
Hopkins, Nancy, 250, 252, 258
Horowitz, Helen, 188
Hosmer, Harriet, 112, 126–29, 132,
135, 136, 154, 205
Hours at Home, 209–10
Howard, Leon, 244–45
Howard College, 172
Howe, Julia Ward: "Battle Hymn of
the Republic" by, 197, 221, 227;
as biographer of MM, 9, 54, 227;
and defense of higher education for
women, 227–28; friendship between
MM and, 227–28; as guest of MM at
Vassar College, 221; and Somerville,
150; travel to Italy by, 119; writings
by, 227
Howland, Sarah, 80
Humboldt, Alexander von, 93
Hunt, Seth, 31
Huxley, Thomas Henry, 173
Hypatia, 148

Institut de France, 19, 64
Irving, Washington, 121
Italian Notebooks (Hawthorne), 245
Italy: American tourists in, 121, 130,
137, 138; artists in, 126–30, 136, 137,
154; astronomers and other scientists
in, 92–93, 116, 133–35, 140, 145–52;
and Byron, 119–20; Florence in,
137–52; Margaret Fuller in, 76, 116,
118–21; Galileo from, 127, 132–33,
134, 137, 138–44; Hawthornes' travel
in, with MM, 121–32, 135–36, 245;
Melville in, 245; MM's travel in,
121–52; nationalism in, 119–20;
observatories in, 133–36, 138, 140;
questions of sex and gender in

Rome, 124–26; Rome in, 118, 119–36, 137, 138; Mary Somerville in, 93, 127, 145–52; Vatican Observatory in, 133–36

Jackson, Andrew, 78
Jefferson, Thomas, 48
Jeffrey, Lord Francis, 178–79
Jenks, Samuel Haynes, 8, 9, 26
Jewett, Milo P., 166–67, 180–82, 187, 235
Johns Hopkins University, 237
Johnson, Samuel, 92, 99
Joule, James Prescott, 114, 151
Journal of the Royal Society, 15
Joy, David, 30–31, 162
Judson Female Institute, 167
Julius Caesar (Shakespeare), 58
Jupiter, 61, 142

Kaestle, Carl, 25
Kalonji, Gretchen, 251
Kelvin, Lord, 114
Kendall, Joshua, 177
Kendall, Phebe Mitchell, 2–4, 11, 84, 110, 154, 199
Kepler, Johannes, 47
Kerber, Linda, 22
King, Mary, 195, 196
Kirch, Maria, 60
Klimasmith, Betsy, 81
Kohlstedt, Sally, 22, 25, 47
Kuhn, Thomas, 257

Ladd-Franklin, Christine, 237–38
Lalande, Jérôme de, 47–48, 60, 62
Lander, Louisa, 126–27, 132, 154
Lankford, John, 78, 203, 240, 249
LaPlace, Pierre-Simon, 145–46, 149
Laqueur, Thomas, 49
Laschever, Sara, 186
Lassell, William, 96, 97, 114
Leibniz, Gottfried, 99–100

Lepaute, Nicole-Reine, 48, 60, 62
Letters from an American Farmer (Crèvecoeur), 79–80
Leverrier, Urbain, 92, 108, 116
Levin, Miriam, 12–13, 170, 176, 202
Lewis, Graceanna, 242–43
Lewis and Clark expedition, 78
libraries: and athenaeum movement, 21, 29–34; Lydia Coleman Mitchell as librarian, 1, 83; MM as librarian of Nantucket Atheneum, ix–x, 29–34, 41, 68, 69, 73, 74, 84, 90, 92, 162; salary of librarians, 34, 69
Livermore, Mary, 221
Liverpool, England, 94–98
London, England, 98–102, 104–5
Loomis, Elias, 32, 70
Lossing, Benson, 198–99
Lunar Society, 101
Lyell, Charles, 101, 114
Lyman, Hannah, 211–12
Lyon, Mary, 38, 168, 170, 172

Mabee, Carolyn, 250
Macaria (Evans), 164–65
Macleish, Martha Hillard, 255
Macy, Alfred, 159
maleness. *See* sexual identity
Manchester, England, 96
Mann, Horace, 27
The Marble Faun (Hawthorne), 123–25, 126, 130, 131–32, 136
Maria Mitchell Association, 250
marriage: companionate marriage of Mitchells, 81; Margaret Fuller on, 77; importance of, for women, 83
Martin, Robert K., 245
Martineau, James, 98
mathematics: and astronomy, xi–xii, 43, 69–70; Caroline Herschel's skills in, 42–43; discovery of calculus, 99–100; and liberal arts in nineteenth century, 37; MM's education

and accomplishments in, 6, 10–11, 32, 34, 54, 69–70; women students of, 15–16, 24

Maury, Antonia, 236–37, 238

Maxwell, James Clerk, xv, 37–38, 99, 101, 114, 248

Melville, Herman, 10, 32, 33, 95–96, 159, 243–47

Memoirs of the American Academy of Arts and Sciences, 22

Mercury, xiii

Methodists, 218

Milton, John: and astronomy, 59, 84, 137, 140, 142, 144; on comets, 59, 60; compared with Barrett Browning, 143–44; Dryden's poem on, 147–48; in Florence, 137; and Galileo, 59, 137, 144; MM on, 59, 140, 141–42; and MM's trip to Europe, 92, 99; writings by, 59, 60, 84, 118, 141

mind-body dualism, 45–46

MIT (Massachusetts Institute of Technology), 237, 250

Mitchell, Andrew, 2, 3, 33, 218

Mitchell, Anne (Annie), 2, 5, 8, 153, 164, 207–8, 214–15

Mitchell, Eliza, 2

Mitchell, Eliza Catherine (Kate). *See* Dame, Kate Mitchell

Mitchell, Francis Macy, 2

Mitchell, Henry, 2, 13, 33–34, 64

Mitchell, Lydia Coleman: children of, 2, 81, 214–15; and children's education, 4–5; death of, 163–64; homes of, 2, 5–7, 31–32, 53, 81–82, 212; illness of, xiii, 89, 153–54, 160–61; marriage of, 2, 81; and Nantucket Philosophical Institution, 19; physical appearance of, 1; as Quaker, 214; reading by, 1–2, 5, 32, 91, 92; as teacher and librarian, 1, 81, 83

Mitchell, Maria: aging of, 240–42; biographies of, 9, 54, 227, 248–49; birth of, 2; death of, 242; diaries of, 73, 82, 89; domestic image of, after her death, 248–49; education of, 10, 27; in England, 91–115; European trip by, 91–152; finances of, 34, 69, 74, 93, 94, 157, 180, 184–87, 189–90; in Florence, Italy, 137–52; in France, 115–19; and friends' deaths, 89, 90; friendships of, with women, 83–90, 197–200, 228; and Hawthornes in Rome, 121–32, 135–36; and historical progress, 141–44; homes of, 5–7, 31–32, 53, 74, 81–82, 164, 187, 212; in Italy, 121–52; as librarian of Nantucket Atheneum, ix–x, 29–34, 41, 68, 69, 73–76, 84, 90, 92, 162; in Lynn, Mass., 164, 187, 241–42; mapping of Nantucket by, 34; and mother's illness, xiii, 89, 153–54, 160–61; name of, 2; in New Orleans, 94; obituaries and memoirs of, 242–43, 248–49; papers of, 189, 199; personality and Quaker sensibilities of, x, xi, xii, 33, 55, 56, 67, 94, 196–97, 198, 228, 239–40, 242; physical appearance and clothing of, x, xii, 33, 94, 106–7, 228; poetry by, 35–37, 40, 91–92, 190–91, 194, 195, 197, 199; portrayal of, in Evans' *Macaria,* 164–65; portrayal of, in Melville's "After the Pleasure Party," 243–47; public speaking by, 84–85, 87, 89–90, 127, 135–36, 165, 185–86, 189, 227, 228–29; and Quaker beliefs, 33, 216, 218; reading by, 32–34, 37, 41, 76, 91, 92, 98, 116, 120, 121, 131, 141; in Rome, Italy, 118, 119–36; as single woman, 72, 83–84; as teacher in her own Nantucket school, 28–29; and Unitarianism, 33, 211–16, 219–23; writings by,

59, 84, 112–13, 125, 127, 149–52, 154, 155–56, 164, 178–79, 206, 209–10; youth and family of, 2–8, 161, 212, 214–15

—as astronomer: astronomical assistant to WM and later collaborative relationship with WM, 5–7, 18–19, 24, 34, 43, 44, 81, 164, 207–8; astronomical interests and studies of MM during youth, 5–7, 9, 10–11, 13–14, 18–20, 24, 27; celebrity status of MM, ix, xii, 57, 70–71, 73–76, 78, 82–83, 239; comet discovery, xvi, 53–58, 63–64, 67, 70–71; honors and awards, xii, xiv, 57, 68, 70, 138, 149, 209, 227, 231; inaccurate portrayal of MM, 254–55; mathematics education and competency, 6, 10–11, 13, 32, 34, 54, 69–70; *Nautical Almanac* employment, xi–xii, 68–70, 74, 90, 93, 94, 155, 157, 163, 180, 189, 200, 201, 231, 254; observatories for, ix–xi, 32, 53, 70, 82, 158–59, 164, 178, 241, 242; telescopes for, x–xi, 97, 157–58, 241; and U.S. Coastal Survey, 66, 70, 208. *See also* Vassar College

Mitchell, Phebe. *See* Kendall, Phebe Mitchell

Mitchell, Sally, 2, 3, 19

Mitchell, William: as astronomer, 3–4, 7–12, 18–19, 24, 42; astronomy lectures by, 7–8, 10, 84, 214; children of, 2, 81, 214–15; and children's education, 4–5, 9–10; college not attended by, 2, 28; death of, 207–8; and discovery of comet by MM, 55, 56; employment of, 4, 5, 7, 11–12, 26, 31, 64, 66; finances of, 5, 6, 11, 31, 74; and Harvard College, 67–68; and Harvard Observatory, 193; homes of, 2, 5–7, 31–32, 53, 74, 81–82, 212; mapping of Nantucket by, 34; marriage of, 2, 81; MM as astronomical assistant to and collaborator with, 5–7, 18–19, 24, 34, 43, 44, 81, 164, 207–8; and Nantucket Philosophical Institution, 19, 20, 24, 25; as Quaker, 214–15; relationship between Walter Folger and, 24–25; retirement of, in Lynn, Mass., 164; as schoolmaster, 4, 7, 11, 26, 27, 81, 212, 216; telescope of, 10, 11, 18, 29; at Vassar College with MM, 188, 192, 207–8

Mitchell, William Forster, 2

Montagu, Elizabeth, 179

Monthly Notices for the Royal Astonomical Society, 56, 57

Mott, Lucretia, xv–xvi, 83, 163

Mount Holyoke Female Seminary, 38, 168, 170, 175–76, 188, 202

Nantucket: architecture of, 160; astronomy as interest in, 7–10, 14; astronomy lectures by WM in, 7–8, 10, 84, 214; bank and business district in, 31, 53, 81–82; Coterie literary club in, 35–37; eclipse (1831) viewed from, 18–19, 208; intellectual culture in, 13–14; libraries in, ix–x, 1, 29–34, 41, 68, 69, 73–76, 84, 90; mapping of, by WM and MM, 34; as maritime community and whaling industry in, 79–80, 94–95, 153, 160, 218; Mitchell family homes in, 2, 5–7, 31–32, 53, 81–82, 212; MM's vacations in, 240; observatories in, ix–xi, 32, 53, 63, 70, 158–59; paintings of, 16; poetry on, 16–17; population and decline of, 98, 153, 217; Quaker meetinghouse in, 217; Quakers in, 4, 8, 13, 14, 15, 25–26, 159–60, 216–18; schools in, 4, 10, 15, 25–29; separate spheres in, 79; summer visitors in, 72, 73, 74–75;

Unitarian Church in, 217, 219–20; winter season in, 72–73; women's challenges in, 79–80; and women's education, 8–9, 13–18, 21, 41, 115

Nantucket Anti-Slavery Convention, 162–63

Nantucket Anti-Slavery Society, 162

Nantucket Atheneum, ix–x, 1, 29–34, 40–41, 68, 69, 73–76, 84, 90, 92, 120, 162

Nantucket Inquirer, 8, 22, 155, 162, 214

Nantucket Inquirer and Mirror, xii

Nantucket Philosophical Institution (NPI), 15, 19, 24, 25, 30, 44, 84

National Academies of Science and Engineering, 250–51

National Council of Women of the United States, 256

natural philosophy, 14, 48–49, 51, 113–14, 169. *See also* science

Nautical Almanac: and Civil War, 163; funding for, 68; MM's employment at, xi–xii, 68–70, 74, 90, 93, 94, 155, 157, 163, 180, 189, 200, 231, 254; MM's resignation from, 201, 254; Neptune's orbit in, xi; reasons for publication of, 68–69; salary paid by, 69, 180; staff of, xi, 69; Venus's orbit in, xi–xii, 69–70, 74, 201, 231, 254

Nautical Survey, U.S., 209

Naval Observatory, U.S., 12, 66, 68

Neeley, Kathryn, xv, 148, 149

Neptune, xi, 106, 108, 116

Newcomb, Simon, 69

Newman, Louise, 174–75

New Orleans, 94

Newton, Sir Isaac: accomplishments of, 99–100; grave of, 100; gravitation theory of, 23; impact of, on religious beliefs, 220; inverse-square law of, 222; mathematics of, 24, 99–100; and MM's trip to Europe, 92, 100–101, 105, 106; and Royal Society, 99, 101; on solar orbits, 62

New York City, 98

New York Times, 240, 253

New York Tribune, 119

nonviolence. *See* pacifism and nonviolence

Norcross, Emily and Louise, 39

Norling, Lisa, 80

NPI. *See* Nantucket Philosophical Institution (NPI)

nudity, attitudes toward, 126, 127, 218

Oberlin College, 168, 172

observatories: J. Q. Adams' support for national observatory, 65–66; in England, 66–67, 96–97, 102–4, 107; in Europe generally, 64, 116; in France, 116; Greenwich, England, observatory, 66–67, 97–98, 102–4; at Harvard, 11, 27, 32, 44, 64–65, 66, 67–68, 103, 193, 236–37; in Italy, 133–36, 138, 140; Maria Mitchell Observatory in Nantucket, 63; of MM in Lynn, Mass., 164, 178, 241, 242; in Russia, 64, 109; U.S. Naval Observatory, 12, 66, 68; at Vassar College, 183, 188–89, 191, 192–93, 203–6, 208, 235, 237; WM's and MM's observatories in Nantucket, ix–xi, 32, 53, 70, 82, 158–59. *See also* astronomy; telescopes

opium, 80

The Origin of Species (Darwin), 102, 144, 172–73

Oxford University, 149

pacifism and nonviolence, 160, 216–17, 218

Paradise Lost (Milton), 59, 84, 141, 143

Paris, France, 116–19

Parker, Theodore, 212, 216, 219, 220

Peabody, Elizabeth, 98, 157

Peabody, Elizabeth Palmer, 214
Peabody, Sophia. *See* Hawthorne,
    Sophia Peabody
Peel, Robert, 149
Philadelphia Athenaeum, 29–30
Philbrick, Nathaniel, 80, 245
Phillips, Wendell, 163
Pickering, Edward, 44
Pierce, Benjamin, xi, 12, 27, 54, 68,
    202
Pierce, Cyrus, 10, 27, 29, 202
Pierce, Edward, 69
planetary motion, xi–xiv, xvii. *See also*
    specific planets
Plutarch, 58
Poe, Edgar Allan, 39, 40, 86
poetry: and astronomy, xvii, 38–39,
    142, 190–91; by Barrett Browning,
    86, 91, 92, 105, 112, 123, 143–44; by
    Byron, 91, 120, 130; by Dickinson,
    38–39, 40, 213; Melville's "After the
    Pleasure Party," 243–47; by Milton,
    59, 60, 84, 118, 141; by MM, 35–37,
    40, 91–92, 190–91, 194, 195, 197,
    199; on MM, 196–97, 231; on Nan-
    tucket, 16–17; by Vassar students,
    196–97, 231, 236, 238; by Whewell,
    146, 148; on women's education,
    17–18, 190–91. *See also specific poets*
Polwhele, Richard, 52
psychology, 237–38
Ptolemy, xiii, 132

Quakers: and abolitionist movement,
    159–61; Discipline and beliefs of,
    4, 8, 32, 159–60, 214, 217–19; and
    education, 24, 25–26; factions of,
    217, 218–19; Benjamin Franklin as,
    28; meetinghouse of, in Nantucket,
    217; Mitchell family as, 212–15;
    MM's Quaker sensibility, 33, 55, 56,
    94, 106–7, 198, 228; MM's ques-
    tioning of beliefs of, 33, 216, 218;

and modesty of Mitchell family
    members, 33, 55, 117–18; in Nan-
    tucket, 4, 8, 13, 14, 15, 25–26,
    159–60, 216–18; and natural philos-
    ophy, 14; and nonviolence and
    pacifism, 160, 216–17, 218; prejudice
    against, 211; Whittier as, 86; and
    women's education, 13, 15, 256

Raphael, 126
Raymond, John H., xvi, 185, 206, 226,
    239
Raymond, Mary, 241
Redburn (Melville), 95–96
religion. *See* Quakers; Unitarianism
Renwick, James, 167, 188
Ricca, Brad, 39
Rich, Adrienne, 248
Richards, Ellen Swallow, 237, 238
Ripley, Mrs., 86
Rome, Italy, 118, 119–36, 137, 138
Rossiter, Margaret, 22, 235, 237
Royal Astronomical Society, 56, 57,
    149
Royal Society: Franklin as member
    of, 15; Huxley on denying women
    membership in, 173; male members
    in, 105, 149; presidents of, 99, 101,
    173; and scientific community, 64;
    women members in, 19, 110, 111,
    138, 149
Rümker, Frau, 54, 60
Runkle, John, xi
Russell, Amelia, 85
Russell, Ida, 83, 85–89, 90
Russell, Jonathan, 85
Russett, Cynthia Eagle, 174, 225, 248
Russia, 64, 93, 109
Rutgers University, 227, 231

Sand, George, 115, 117
San Francisco Chronicle, 252
Saturn, 142

Scarpellini, Caterina, 60

Schiebinger, Londa, 173

science: Baconian scientific method, 43; coinage of word "scientist," xv, 105, 145, 146–47; collaboration and teamwork in, 43–44, 47; dynamism and social rigidity of, in nineteenth century, 173–74, 248; Emerson on, 40; exclusion of and barriers to women in, xiv–xv, 45, 225, 230, 234–38, 248–59; and humanities in nineteenth century, 37–41; and industrialization, 173; masculiniza- tion of, 45–46, 173–74, 175, 225, 257, 258; MM on scientists as seekers of truth, 222–23; and natural philos- ophy, 14, 48–49, 51, 113–14, 169; professionalization of, 156–57, 173–74, 175; publishing in, 93; stereotype of, as pursuit of solitary men, 44–45; Summers on women's unfitness for, 250, 252–53; unifica- tion of nineteenth-century sciences, 146–49; and Unitarianism, 213–14, 220; in U.S. in eighteenth century, 48–49; women assistants in, 44, 45, 236–37, 257; women members of scientific institutions, 19–20, 68, 70, 110, 138, 149; women professors in, 183, 253; women scientists and women students of, xiv–xviii, 12–18, 40, 46–48, 83, 147–52, 169, 247–48, 254, 255–56, 258–59; women scien- tists in twenty-first century, 250–54, 258; in women's colleges, 175–76; women's unique contribution to, xv, 147, 152, 210, 229–30, 249. *See also* astronomy; England; Europe; France; Italy; *and specific scientists*

Scientific American, x, 158, 206, 256

Scotland, 105

Scott, Mary A., 193–94, 197–98

Scribner's, 225

Secchi, Father Angelo, 54, 93, 133–35

Seneca Falls Convention (1848), xvi, 57

Sewall, Richard, 39

Sex and Education: A Reply to E. H. Clarke's Sex in Education (Howe), 227

Sex in Education (Clarke), 224–30

sexual identity: in Age of Reason, 50–52; analogous relationship between women as inferior sex and inferior races, 174–75; and changing meanings of Urania and Uranians, 243–47; Descartes' one-sex model of, 45–46, 49; domestic ideology regarding women, xii–xiii, 125, 156, 178, 179–80, 187; Margaret Fuller on, 77, 244; in late nineteenth cen- tury, xii–xiii, 49; and masculiniza- tion of science, 45–46, 173–74, 175, 225, 257, 258; in Melville's "After the Pleasure Party," 243–47; and MM's article on solar eclipse, 209–10; negative connotations of word "female," 170–73, 175, 176; physiological basis for "true sex," 225; private versus public sphere, 78–79; race and gender ideology, 163; in seventeeth and eighteenth centuries, 49–50; and true woman- hood, 187

Shakespeare, William, 58–59, 92, 99

Shalala, Donna, 250–54

Shapley, Harlow, 44, 45

Shelley, Mary, 51, 60

Shoemaker, Carolyn, 61

Shoemaker-Levy 9 comet, 61

Silliman, Benjamin, 92

Silliman's Journal, 56, 156, 164, 206

slavery, 94, 156, 159–64, 171–72. *See also* abolitionist movement

Smith College, 176, 188, 224, 231

Smithsonian Institution, 68, 70, 92, 93, 167, 208

Smyth, Annarella, 102, 114

Smyth, Admiral William Henry, 102, 114, 159

Snow, C. P., 37, 257

Society of Friends. *See* Quakers

Solomon, Barbara, 168

Somerville, Mary: bust of, in Vassar Observatory, 193; coinage of word "scientist" referring to, xv, 105, 145, 146–49; compared with Caroline Herschel, 112; death of, 149; femininity of, 149–50; honors and awards for, 149; and LaPlace's astronomy, 145–46, 149; MM's acquaintance with, in Italy, 93, 116, 127, 145, 150–52, 174; MM's article on, 112, 125, 127, 149–52, 154, 155, 178–79; MM's stories on, to Vassar students, 205; Mott on, xvi; pension for, 149; as role model for MM, 110, 145, 151–52, 156, 241; and Royal Society, 110, 138, 149; as scientist and mathematician, xv, 105, 116, 145–52; Whewell's poetry on, 146, 148; writings by, xv, 145, 146–49, 151

Spelman College, 172

Spring, Rebecca and Marcus, 121

Stanton, Elizabeth Cady, 221

Starbuck, Alexander, 26

stars, photograph of, 103

Stepan, Nancy Leys, 174

Stevenson, Polly, 14, 48

Stillingfleet, Benjamin, 179

Stowe, Harriet Beecher, 116, 117, 118

Struve, Wilhelm, 93, 109, 114

Summer on the Lakes (Fuller), 76

Summers, Lawrence, 250, 252, 258

Swift, Prudence, 94, 109–10, 117

Taylor, John, 96–97

Taylor, John M., 242

telescopes: of Bond family, 47; for discovery of comets, 60–61, 64; in England, 96–98; of Folger, 22; function of, 62–63; and Galileo, 62–63, 142; at Harvard Observatory, 66; in Melville's "After the Pleasure Party," 245–46; MM's telescopes, x–xi, 97, 157–58, 241; neglected telescopes, 97–98; at Vassar Observatory, 189, 204; at Vatican Observatory, 134; WM's telescope, 10, 11, 18, 29, 33, 97. *See also* astronomy; observatories

Thoreau, Henry David, 195

Tolley, Kim, 12–13, 169

Tougaloo College, 172

Toynbee, Arnold, 107

transcendentalism, 215

Trollope, Anthony, 91, 121

Troy Female Seminary, 165, 168, 177

true womanhood, 187

Truth, Sojourner, 33, 163

Unitarianism, 33, 98, 160, 211–16, 219–23

United States magazine, 89–90

universities. *See* colleges and universities; education; women's education; *and specific universities*

University of Iowa, 168

University of Santa Cruz, 251

University of Wisconsin, 168

Urania: changing meanings of, 243–47, 248; in Melville's "After the Pleasure Party," 243–47; as muse of astronomy, 47, 100, 116, 243

Uranus, 42, 110, 243

U.S. Coastal Survey, 11–12, 27, 56, 64, 66, 70, 208

U.S. Nautical Survey, 209

U.S. Naval Observatory, 12, 66, 68

Vassar, Matthew: and founding of Vassar College, 166–68; on furniture for professors' dwellings, 187–88; and hiring of MM, 177, 178, 187;

and hiring of women professors, 177–78, 182–84, 187; and naming of Vassar College, 170–72; and observatory at Vassar College, 189

Vassar College: architecture of, 167, 188–89; curriculum of, 170, 183; and defense by MM of higher education for women, 226–30; and eclipse expedition (1869) to Iowa, 208–10; end-of-the-year celebrations of MM with students at, 194–96, 241; exclusion of black women by, 172, 175–76, 233–34; faculty publications at, 206; founding of, 166–68; grading by MM at, 194; graduates of, 207, 233, 234–38, 257; hiring of MM by, 165, 177–87; hiring of women professors at, 177, 180–87, 254; housing for male professors at, 189; intimate friendship between MM and her students at, 197–200; Jewett as president of, 180–82, 187; medical records of students at, 227; MM as professor of astronomy at, xv, xvii, 176, 194, 195–96, 200–209, 241; MM's advanced astronomy class at, 201, 207, 208–9, 235; MM's astronomical research at, 200, 203, 206; and MM's priority on women's intellectual culture, 203–7, 226–30, 249, 259; MM's relationship with administration at, 206, 239; MM's retirement from, 241; MM's room at, 188–92, 230–31, 240, 248–49; MM's writing during her professorship at, 206; naming of, 170–72, 175, 176; observatory at, 183, 188–89, 191, 192–93, 203–6, 208, 235, 237; opening of, 192, 224; and opposition to higher education for women, 234–38; pedagogy of MM at, 202–4; Raymond as president of,

xvi, 185, 206, 226, 239; salaries at, 184–87, 189–90, 253; science courses in, 175–76; students of MM at, 176, 193–200, 234–38, 240, 249, 259; students' song honoring MM at, 196–97; Sunday night socializing with MM and students at, 193–94; and Unitarianism of MM, 211–12, 220–23; visitors to and speakers at, 221; WM at, 188, 192–93, 207–8; work schedule of MM at, 200–201, 230

Vatican Observatory, 133–36

Venus (planet): brightness of, xi; as female planet, xiii–xiv, xvii; MM's calculation of orbit of, for *Nautical Almanac,* xi–xii, 69–70, 74, 155, 201, 231, 254; orbit of, xi–xiv; position of, xiii

Venus (sculpture), 126

Vesey, Mrs., 179

Vesta (asteroid), 2

Vesta and vestal virgins, 2

A Vindication of the Rights of Women (Wollstonecraft), 50–51

Virgil, 121, 148

Walker, Sears Cook, xi, 108

Washington, George, 78

Washington, George (African American), 161

Webster, Daniel, 78

Wellesley College, 176, 224

Wendt, Amy, 250

whaling industry, 79–80, 153, 160, 218

Whewell, William: on Barrett Browning's poetry, 105; coinage of word "scientist" referring to Mary Somerville, xv, 105, 145, 146–49; and inductive science, 113; MM's acquaintance with, 92, 105–6,

107, 114, 145; poetry by, 146, 148;
and Royal Society, 149; writings by,
107
Whipple, John Adams, 103
Whitman, Sarah Helen, 86
Whitman, Walt, 39–40
Whitney, Mary, 222, 235, 237, 242
Whittier, John Greenleaf, 86, 88,
241
Willard, Emma, 168, 170
Williams, Lizzie, 212
William the Conqueror, 58
Wollstonecraft, Mary, 50–52, 60
Woman in the Nineteenth Century
(Fuller), 76, 77, 82, 123
The Woman's Journal, ix
women: in abolitionist movement,
156; analogous relationship between
women as inferior sex and inferior
races, 174–75; as Angel in the
House, xii–xiii; as assistants in
science, 44, 45, 236–37, 257; as
astronomers, xvi, 3–4, 5, 42–43, 44,
47–48, 59–61, 83, 110–14; benevolent womanhood, 83; as bluestockings, 178–79; domestic ideology
regarding, xii–xiii, 125, 156, 178,
179–80, 817; employment of, 83,
156; exclusion of and barriers to
women in science, xiv–xv, 45, 225,
230, 234–38, 248–59; Margaret
Fuller on rule of, 75–77, 82; intimate friendships between, 198;
marriage as important for, 83; as
members of Nantucket Philosophical
Institution, 15, 19; as members of
scientific institutions, 19–20, 68, 70,
110, 138, 149; MM on role of gifted
women, 112–13, 140, 229–30, 249,
259; MM's priority on intellectual
culture of, 203–7, 226–30, 249, 259;
negative connotations of word

"female," 170–73, 175, 176; political
rights for, xv–xvi; private sphere for,
78–79; as public speakers, 82–85, 87,
89–90; as scientists and students of
science, xiv, xv–xvii, xviii, 12–18,
40, 46–48, 147–52, 155–56, 169,
247–48, 254, 258–59; as scientists in
twenty-first century, 250–54, 258; as
sculptors, 126–29; single women, 72,
83–84; submission of, 77–78; Summers on women's unfitness for science, 250, 252–53, 258; as teachers,
25–27; true womanhood cult, 187,
225; unique contribution by, to science, xv, 147, 152, 210, 229–30, 249.
See also sexual identity; women's
education; women's rights; *and
specific women*
Women's Congress, 227, 228–29
women's education: in colleges and
universities, 135–36, 166–68, 170,
175–76; dangers of college-level
education for women, 224–28, 232,
233; disciplines not acceptable for,
xvi; and Maria Edgeworth, 51–52;
and employment rights, 232; in
England, 21, 115, 149; expansion and
contraction of generally, xiv, 232,
254; and hiring of women college
professors, 177, 180–87; MM's
defense of higher education for
women, 226–30; MM's priority on
women's intellectual culture, 203–7,
226–30, 249, 259; in Nantucket,
8–9, 13–18, 21, 41, 115; opposition
to women's study of science, 234–38;
poetry on, 17–18, 190–91; and
Quakers, 13, 15, 256; in sciences,
12–18, 175–76, 202, 254, 255–56;
in U.S. generally, 115. *See also* Vassar
College; women's rights; *and specific
schools*

women's rights: and employment rights, 232; Horatio Hale on, 181–82; and racial equality, 232–34; Seneca Falls Convention (1848) on, xvi, 57; and Unitarianism, 216, 220; and vote for women, 233; and white women, 174–75, 233. *See also* women's education

Wood, Frances, 206, 221
Woolf, Virginia, 106
Wright, Helen, 34, 118, 215, 249

Yale University, 28, 32, 70, 92, 93, 169–70, 183

Zenobia in Chains (Hosmer), 127, 128